Ziad Soufi

Phytoplasma in Sugarcane

Ziad Soufi

Phytoplasma in Sugarcane

Molecular Analysis of Phytoplasmas in Sugarcane Plants

Südwestdeutscher Verlag für Hochschulschriften

Impressum / Imprint
Bibliografische Information der Deutschen Nationalbibliothek: Die Deutsche Nationalbibliothek verzeichnet diese Publikation in der Deutschen Nationalbibliografie; detaillierte bibliografische Daten sind im Internet über http://dnb.d-nb.de abrufbar.
Alle in diesem Buch genannten Marken und Produktnamen unterliegen warenzeichen-, marken- oder patentrechtlichem Schutz bzw. sind Warenzeichen oder eingetragene Warenzeichen der jeweiligen Inhaber. Die Wiedergabe von Marken, Produktnamen, Gebrauchsnamen, Handelsnamen, Warenbezeichnungen u.s.w. in diesem Werk berechtigt auch ohne besondere Kennzeichnung nicht zu der Annahme, dass solche Namen im Sinne der Warenzeichen- und Markenschutzgesetzgebung als frei zu betrachten wären und daher von jedermann benutzt werden dürften.

Bibliographic information published by the Deutsche Nationalbibliothek: The Deutsche Nationalbibliothek lists this publication in the Deutsche Nationalbibliografie; detailed bibliographic data are available in the Internet at http://dnb.d-nb.de.
Any brand names and product names mentioned in this book are subject to trademark, brand or patent protection and are trademarks or registered trademarks of their respective holders. The use of brand names, product names, common names, trade names, product descriptions etc. even without a particular marking in this works is in no way to be construed to mean that such names may be regarded as unrestricted in respect of trademark and brand protection legislation and could thus be used by anyone.

Coverbild / Cover image: www.ingimage.com

Verlag / Publisher:
Südwestdeutscher Verlag für Hochschulschriften
ist ein Imprint der / is a trademark of
OmniScriptum GmbH & Co. KG
Heinrich-Böcking-Str. 6-8, 66121 Saarbrücken, Deutschland / Germany
Email: info@svh-verlag.de

Herstellung: siehe letzte Seite /
Printed at: see last page
ISBN: 978-3-8381-3918-0

Zugl. / Approved by: Bayreuth, Bio, Diss, 2012

Copyright © 2014 OmniScriptum GmbH & Co. KG
Alle Rechte vorbehalten. / All rights reserved. Saarbrücken 2014

Contents

1. Introduction 1
1.1. Plant diseases 1
1.2. Phytoplasmas as plant pathogens 2
1.3. Definition of phytoplasmas 2
1.4. Transmission and spread of phytoplasmal diseases 6
1.4.1. Host cycle of phytoplasmas 6
1.4.2. Host Specificity of Phytoplasmas 8
1.5. Phylogenetic position of phytoplasmas 8
1.6. Management and control of phytoplasma diseases 9
1.6.1. Prevention strategy 9
1.6.2. Control of insect vectors 10
1.6.3. Management strategy 10
1.7. Anatomy of phloem cells 10
1.8. Some phytoplasma diseases of sugarcane 11
1.8.1. Sugarcane yellow leaf syndrome 11
1.8.2. Sugarcane white leaf and sugarcane grassy shoot 12
1.9. Detection of sugarcane phytoplasma infections 13
1.10. Sugarcane yellow leaf syndrome in Hawaii 14

2. Material and Methods 17
2.1. Material 17
2.1.2. Chemicals and Enzymes 17
2.1.2.1. Chemicals 17
2.1.2.2. Enzymes 18
2.1.3. Buffers, Solutions 18
2.1.3.1. Buffer and solutions for DNA extraction 18
2.1.3.2. Buffer for gel electrophoresis 19
2.1.3.3. Buffer for PCR 19
2.1.3.4. Buffer for restriction enzymes 19
2.1.3.5. Buffer for polyacrylamide gel electrophoresis 19
2.1.4. Kits 20
2.1.4.1. Isolation of Nucleic Acids for PCR 20
2.1.4.2. Nucleic acids purification 20

- 2.1.4.3. Q-PCR 20
- 2.1.5. Oligonucleotides 21
- 2.1.6. Software for Gene analysis 21
- 2.2. Methods 22
  - 2.2.1. Plant material 22
  - 2.2.2. DNA extraction strategies 22
  - 2.2.3. Polymerase Chain Reaction (PCR) for the detection of phytoplasmas 24
    - 2.2.3.1. Definition of Nested PCR 24
    - 2.2.3.2. Nested PCR Reaction 25
    - 2.2.3.3. First round of PCR 26
    - 2.2.3.4. Nested round of PCR 26
    - 2.2.3.5. Nested-PCR assay (I) 26
    - 2.2.3.6. Nested-PCR assay (II) 27
    - 2.2.3.7. Nested-PCR assay (III) 27
    - 2.2.3.8. Nested-PCR assay (IV) 28
  - 2.2.4. Agarose Gel Electrophoresis 29
  - 2.2.5. Digestion of nested-PCR products 30
    - 2.2.5.1. Inactivation of restriction enzymes 30
  - 2.2.6. Polyacrylamide Gel Electrophoresis 31
    - 2.2.6.1. Steps of operation 31
    - 2.2.6.2. Special equipment 31
    - 2.2.6.3. Detection of DNA in polyacrylamide gels by staining 31
  - 2.2.7. Sequencing and phylogenetic analysis of ribosomal DNA 31
    - 2.2.7.1. Sample Preparation for Value Read Service in Tubes 32
  - 2.2.8. Hot water treatment 32
    - 2.2.8.1. Preparation of the plant material prior to the hot water treatment 33
  - 2.2.9. Sugarcane aphid transmission test 33
    - 2.2.9.1. Insect rearing 33
    - 2.2.9.2. Plant material 33
    - 2.2.9.3. Transmission tests 34
  - 2.2.10. Q-PCR (real-time PCR) assay 34
    - 2.2.10.1. Methods of monitoring DNA amplification in qPCR 34
    - 2.2.10.2. Detection of phytoplasma based on TaqMan qPCR assays 36
  - 2.2.11. Transmission Electron Microscopy (TEM) 37

2.2.11.1. Preparation of thin sections 38

3. Results 39

3.1. Establishment of the test for phytoplasma 39

3.1.1. PCR for detection of phytoplasma 39

3.1.2. Sources of sugarcane samples 41

3.2. Phytoplasma in sugarcane in Hawaii, Cuba, Egypt and Syria 42

3.2.1. Phytoplasma detection by nested-PCR assay (I) and identification by RFLP 42

3.2.2. Phytoplasma detection by nested-PCR assay (II) and identification by RFLP 46

3.2.3. Phytoplasma detection by nested-PCR assay (III) 48

3.2.4. Phytoplasma detection by nested-PCR assay (IV) and identification by RFLP 49

3.3. Phytoplasma in sugarcane in Hawaiian plantations (2009) 52

3.3.1. Phytoplasma detection and identification 54

3.4. Phytoplasma in sugarcane in Hawaiian former plantation fields (2009) 55

3.4.1. Phytoplasma detection and identification 56

3.5. Phytoplasma in sugarcane in Hawaiian breeding station (2010) 57

3.5.1. Phytoplasma detection and identification 57

3.6. Phytoplasma in sugarcane in Hawaiian breeding station and plantations (2011) 58

3.6.1. Phytoplasma in sugarcane in Hawaiian plantation 58

3.6.1.1. Phytoplasma detection and identification 59

3.6.2. Phytoplasma in sugarcane in Hawaiian breeding station (2011) 61

3.6.2.1. Phytoplasma detection by nested-PCR assay (II) and identification 61

3.6.2.2. Phytoplasma detection by nested-PCR assay (III) and (IV) and identification 63

3.6.3. Phytoplasma in sugarcane in different areas close to former plantations 67

3.6.3.1. Phytoplasma detection and identification 67

3.6.4. Phytoplasma in grass weeds in Sumida water cress farm 69

3.7. Phytoplasma in sugarcane in Thailand (2010 – 2011) 70

3.7.1. Phytoplasma in sugarcane in Bang Phra and Khon Kean provinces (2010) 70

3.7.1.1. Phytoplasma detection and identification 71

3.7.2. Phytoplasma in sugarcane in Suphan Buri province (2011) 75

3.7.2.1. Phytoplasma detection and identification 75

3.8. Establishment of TaqMan qPCR assay as another test for phytoplasma 76

3.8.1. Performance characteristics of qPCR 76

3.8.1.1. Efficiency Measurement 77

3.8.1.2. Artificial samples to test sensitivity of qPCR assay 81

    3.8.2. Q-PCR results of the sugarcane samples from different sources 83

    3.8.3. Distribution of phytoplasma in sugarcane 88

  3.10. Hot water treatment in order to get phytoplasma free sugarcane plant 96

    3.10.1. Hot water treatment according to Australian recipe 96

    3.10.2. Hot water treatment with various duration 97

  3.11. Insecticide treatment of phytoplasma-infected sugarcane plant 99

  3.12. Transmission test with sugarcane aphid (*Melanaphis sacchari*) 100

  3.13. Transmission electron microscopy for cytological location of phytoplasma 101

    3.13.1. Anatomy of leaf phloem tissue 101

    3.13.2. Localization of phytoplasma infection 102

    3.13.3. Phytoplasma size and shape 106

    3.13.4. Ultrastructural changes of the phytoplasma infection on leave anatomy 108

4. Discussion 110

  4.1. Establishment of the test for phytoplasma 110

    4.1.1. Efficiency of PCR amplification 110

    4.1.2. Carry-over contamination problems 112

    4.1.3. Q-PCR (real-time PCR) 112

  4.2. Phytoplasma detection and identification by RFLP analysis 115

    4.2.1. Phytoplasma types in Hawaiian and Cuban sugarcane 115

    4.2.2. Phytoplasma types in Thai sugarcane 117

    4.2.3. Phytoplasma types in Egyptian and Syrian sugarcane 118

  4.3. Identification of the phytoplasma strains in sugarcane by phylogentic analysis 120

  4.4. Hot water treatment in order to get phytoplasma free plant 121

  4.5. Transmission test with sugarcane aphid 122

  4.6. Transmission electron microscopy for cytological location of phytoplasma 123

5. Summary 125

6. Zusammenfassung 127

7. Acknowledgement 129

8. References 130

9. List of figures 145

10. List of tables 148

Abbreviations

AAY	American aster yellows
AY	aster yellows
bp	base pair
BSA	bovine serum albumin
°C	degree Celsius
Ct	threshold cycle
CTAB	hexadecyltrimethylammonium bromide
dd	double distilled
dNTP	deoxyribonucleotide triphosphate
DNA	deoxyribonucleic acid
EDTA	Ethylenediaminetetraacetic acid
FAM	6-carboxyfluorescein
Fig	figure
GS	grassy shoot
HWT	Hot water treatment
kb	kilo base pairs
l	liter
µ	micro
n-PCR	nested- polymerase chain reaction
OY	onion yellows
PCR	polymerase chain reaction
PVP	polyvinylpyrrolidone
Q-PCR	Quantitative PCR
qPCR	Quantitative PCR
rRNA	ribosomal RNA
RFLP	restriction fragment length polymorphism
RNase	ribonuclease
rRNA	ribosomal RNA

rpm	revolutions per minute
RT	room temperature
RYD	rice yellow dwarf
S	second
SCYLP	sugarcane yellow leaf phytoplasma
SCYLV	sugarcane yellow leaf virus
SCWL	sugarcane white leaf
SCGS	sugarcane grassy shoots
SGS	sorghum grassy shoots
TAMRA	6-tetramethylrhodamine
TBE	Tris base- Boric acid- EDTA
TEM	Transmission Electron Microscopy
tRNA	Transfer RNA
U	unit
UV	Ultraviolet
YLS	yellow leaf syndrom

1. Introduction

1.1. Plant diseases

"Plants make up the majority of the earth's living environment as trees, grass, flowers, etc. Directly or indirectly, plants also make up all the food on which humans and all animals depend. Plants are the only higher organisms that can convert the energy of sunlight into stored, usable chemical energy in carbohydrates, proteins, and fats" (Agrios, 2004).

"Plant diseases are very important part of plant protection within the system of plant/crop production. Diseases can significantly lessen the growth and yield or reduce the utility of a plant or plant product. Healthy plants grow and function to the maximum of their genetic potential. However, plants are considered to be diseased when they are negatively affected by a disease-causing agent that lead to interfering with their normal development and physiological functions" (Agrios, 2004).

"Correct diagnosis of the cause of a disease is an essential step in order to construct a convenient strategy to manage the plant disease. Usually, the first step includes the determination of the probable cause of the disease: whether the disease is caused by an infectious agent (pathogen) or environmental factor. Since diseases in plants are caused by either non-living (abiotic, non-parasitic, non-infectious, 'non-pathogenic') environmental factors or living (biotic, parasitic, pathogenic, infectious) agents. On the other hand, plant diseases are grouped based on the causal agent involved (deficiency diseases, fungal diseases, bacterial diseases, viral diseases, mollicutes diseases, etc.), the plant part affected or the type of symptoms" (Agrios, 2004).

In general, plant disease is any growth or developmental condition that is not "normal" to that plant and can usually diminish its economic or aesthetic value. In many cases, the plant pathologists depend on symptoms and signs of the disease for hypothetical diagnosis of diseases in plants. The characteristic internal or external alterations showed by the plant in reaction to the disease-causing agent are called symptoms.

Plant pathology is the study of the pathogens and of the environmental factors that cause disease in plants, and the methods of preventing or controlling disease and reducing the damage it causes. Uncontrolled plant diseases may result in less food and higher food prices, or in low-quality food (Agrios, 2004). Over the last decades scientists in molecular plant pathology have also established a new set of diagnostic tools and techniques that are used to detect and identify pathogens even when they are present in diminishing small numbers or in mixtures with closely related pathogens. Such tools include detection with monoclonal antibodies, calculation of percentages of hybridization of their nucleic acids, and determination of nucleotide sequences of the nucleic acids of the pathogens. Since the mid-1980s, decisive DNA fragments, so called DNA probes, complementary to specific segments of the nucleic acid of the microorganisms, have been labelled with radioactive isotopes or fluorophores and are used extensively for the detection and identification of plant pathogens (Agrios, 1997).

1.2. Phytoplasmas as plant pathogens

For nearly 70 years after viruses were discovered, many plant diseases were described that showed symptoms of general yellowing or reddening of the plant, or of shoots proliferating and forming structures that resembled witches brooms respectively. These diseases were thought to be caused by viruses, but no viruses could be found in such plants (Agrios, 2004).

In 1967, Japanese scientists found out those plant pathogens known recently as phytoplasmas were the potential causes of plant yellows disease (Doi et al., 1967). After detection of these pathogens which lacking cell wall in the phloem tissue of infected plants and the evidence presented that these microorganisms, rather than hypothetical, undetectable viruses were the causal agents, plant pathologists and entomologists started to re-investigate of many plant diseases that previously expected as virus diseases since these causal agents could not be cultured in artificial media like the viruses. In the following years, many studies indicated the association of these microorganisms that previously named mycoplasma-like organisms with many different plant diseases. The methods followed for phytoplasms detection were electron microscopy of thin sections of the phloem, and tetracycline treatment of diseased plants as described by the Japanese plant pathologists.

1.3. Definition of phytoplasmas

Bacteria and mollicutes are prokaryotes. These are single-celled microorganisms whose genetic material (DNA) is not bound by membrane and therefore is not organized into a nucleus. Their cells consist of cytoplasm containing DNA and small ribosomes (70S). The cytoplasm in mollicutes is surrounded by cell membrane only, but in bacteria it is surrounded by a cell membrane and a cell wall (Agrios, 2004). Plant mollicutes also grow in the

Introduction

alimentary canal, hemolymph, salivary glands, and intracellularly in various body organs of their insect vectors.

Phytoplasmas (previously termed mycoplasma like organisms) belong to the class *Mollicutes*. They are single-celled plant pathogenic sub-microscopic microorganisms and similar to bacteria but much smaller than others (with a diameter less than 1 µm) (Figure.1.1). Since they lack cell wall, they can change shape (pleomorphic organisms). Phytoplasmas exist in phloem sieve elements in infected plants (Doi et al., 1967; Whitcomb and Tully, 1989) (Figure.1.2).

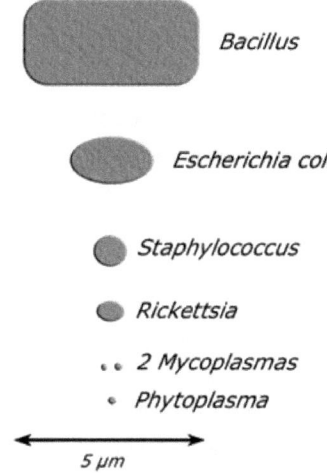

Figure.1.1. Comparison of sizes of some eubacteria. Phytoplasmas (previously termed mycoplasma like organisms) belong to the class *Mollicutes*. They are single-celled plant pathogenic sub-microscopic microorganisms and similar to bacteria but much smaller than others (with a diameter less than 1 µm). Phytoplasmas have the same size as mycoplasmas. Figure taken from Saskias phytoplasma website,http://www.jic.ac.uk/staff/saskia-hogenhout/index.htm.

3

Introduction

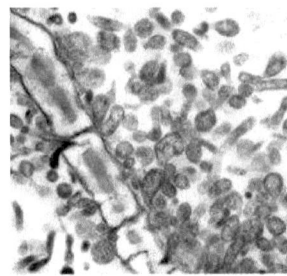

Figure.1.2. Pleomorphic phytoplasmas in sieve tubes. Phytoplasmas multiply in the sieve tubes of phloem and circulate through the sieve pores. Figure taken from INRA website http://www.international.inra.fr/research/some_examples/a_national_collection_of_phloem_bacteria.

Phytoplasmas grow and reproduce only in living host tissue. Therefore, unlike most human and animal mycoplasmas, phytoplasma cannot be cultured on artificial nutrient media (cell-free media), involving the media on which all typical mycoplasmas grow (Lee and Davis, 1992). This inability has made it difficult to determine the taxonomic status of phytoplasmas by the traditional methods applied to cultured prokaryotes. Currently, because the development of molecular tools has made it possible to identify the phytoplasma based on the nucleotide sequence of the 16S rRNA gene (Gundersen et al., 1994; Limand Sears, 1989; Namba et al., 1993; Sawayanagi et al., 1999; Seemüller et al., 1994), since these phytopathogenic mollicutes are uncultivable and experimentally inaccessible in their hosts, knowledge of their biological properties is also restricted (Christensen et al., 2005). Therefore, the mechanisms by which phytoplasmas cause plant diseases are not well understood and make it difficult to develop means to control them. Phytoplasmas, however, have been extracted from their host plants and from their vectors in more or less pure form, and for most of them antisera, including monoclonal antibodies, have been prepared.

Specific antibodies, DNA probes, RFLP profiles, and analysis of 16S rRNA genes have become extremely useful in the detection and identification of the pathogen in suspected hosts, in grouping and classifying the pathogens, and in controlling these diseases through production of pathogen-free propagating stock (Agrios, 2004). Recently, moreover, the full genomes of some phytoplasmas have been sequenced (Bai et al., 2006; Oshima et al., 2004) allowing new insights into their requirements (Christensen et al., 2005).

Introduction

Molecular-based tools, and sensitive detection procedures developed in the past decade have permitted great advances in the diagnostics of diseases caused by phytoplasmas and have facilitated the characterization of phytoplasmas (Davies and Clark, 1992; Firrao et al., 1996).

Plants infected with phytoplasmas show many different symptoms involve unseasonal yellowing, reddening or discolorations of the leaves, shortening of the internodes with stunted growth, smaller leaves, and excessive proliferation of shoots resulting in witch's broom(a dense mass of shoots grows from a single point, with the resulting structure resembling a broom or a bird's nest), phyllody (the development of floral parts into leafy structures),sterility of flowers, necrosis of the phloem tissues and the general decline and the death of the plant (Agrios, 1997; Kirkpatrick, 1992; McCoy et al., 1989).

It seems that certain effects are on individual cells, while others are on cell interactions. The striking morphological and metabolic changes suggest that toxins might be produced by the microorganisms that influence the hormonal balance and interfere with photosynthesis.

To date, these unique plant pathogens have been associated with diseases in several hundred plant species covering a geographic range from temperate to tropical areas and including many important food, vegetable and fruit crops; ornamental plants; timber and shade trees (McCoy et al., 1989; Sinclair et al., 1994; Lee et al., 2000) (Figure.1.3). Furthermore the list of diseases caused by phytoplasmas continues to grow. Phytoplasma, especially sugarcane white leaf phytoplasma, are responsible for losses of over 100 million baht (about Australian $4.5 million) each year to the sugarcane industry in Thailand (Wongkaew et al. 1997) and in India phytoplasma is also emerging as a major problem for sugarcane. Important plant diseases caused by phytoplasmas are aster yellows, apple proliferation, coconut lethal yellowing, elm yellows, peach X-disease, big bud diseases of solanaceous plants, and many more.

Introduction

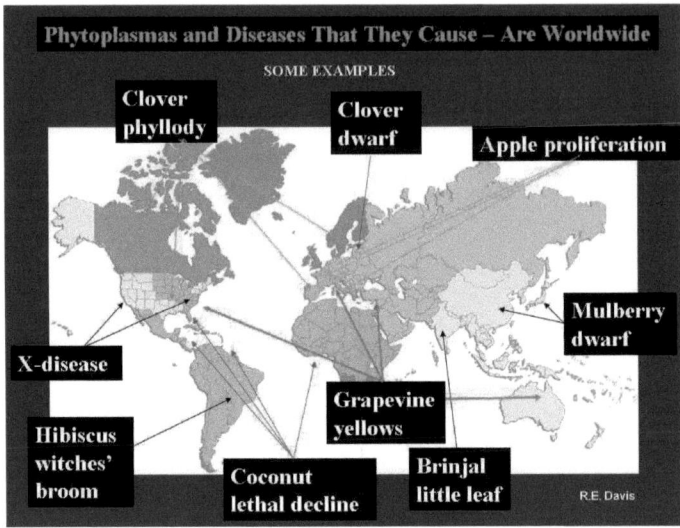

Figure.1.3. Phytoplasmas and their diseases are worldwide. Phytoplasmas have been associated with diseases in several hundred plant species covering a geographic range from temperate to tropical areas and including many important food, vegetable and fruit crops; ornamental plants; timber and shade trees. Figure taken from http://plantpathology.ba.ars.usda.gov/pclass/pclass_phytoplasma_spread.html.

1.4. Transmission and spread of phytoplasmal diseases

Phytoplasmas are phloem-limited plant pathogens that are invading primarily sieve tube elements. Phytoplasmal diseases are spread primarily by sap-sucking insect vectors, most commonly leafhoppers but also some psyllids and plant hoppers (Weintraub and Beanland, 2006; Ploaie, 1981) and all these vectors are belonging to the Hemiptera (Rhynchota), which are a large and diverse order of exopterygote insects, which occur throughout the world and there are more than 60.000 species in around 100 families. The Hemiptera is now divided into 3 sub-orders: Heteroptera (true bugs), Sternorrhyncha (scaleinsects, aphids, whiteflies, psyllids) and Auchenorrhyncha (leafhoppers, planthoppers, cicadas, treehoppers, and spittlebugs).

1.4.1. Host cycle of phytoplasmas

An insect vector acquires the phytoplasma after feeding on an infected plant for several hours or days (acquisition feeding). For 10 - 45 days, the phytoplasma moves through the insect and multiplies within specific organs (Ammar and Hogenhout, 2006). After this incubation period the insect is able to transmit the phytoplasma to uninfected plants when it feeds (Murral et al., 1996). Next, multiplication and spread of phytoplasmas in the host plant is accompanied by the appearance of disease symptoms (Figure.1.4). An infected insect will be able to spread disease for the rest of its life.

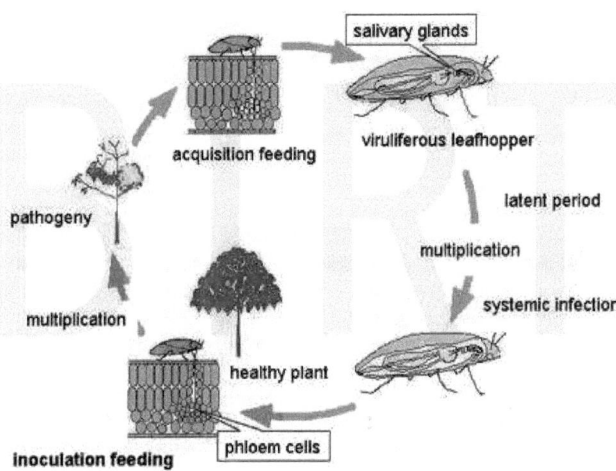

Figure.1.4. Host cycle of phytoplasmas.(Agrios, 1997).

Phytoplasma diseases tend to occur more often in outdoor planting than in greenhouse, where it is easier to detect and control leafhoppers. Phytoplasmas can be spread by vegetative propagation through cutting, storage tubers, rhizomes, or bulbs (Lee and Davis, 1992). In commercial plantations, infections can be obtained also by grafting. However, phytoplasma cannot be transmitted mechanically by inoculation with phytoplasma-containing sap. Furthermore, phytoplasmas are not known to be transmitted through seed or pollen.

1.4.2. Host Specificity of Phytoplasmas

Plant host range for a phytoplasma is dependent upon vectors specificity and feeding habits (behaviours) (monophagous, oligophagous, and polyphagous) of these vectors. For example, North American aster yellows phytoplasmas (16SrI-A,-B) were transmitted experimentally by the polyphagous leafhopper Macrosteles fascifrons and other vectors to 191 plant species belonging to 42 families (McCoy et al., 1989). Not all vectoring insects can transmit all phytoplasmas and there are specific interactions of a particular phytoplasma with its insect vector. Some phytoplasmas, such as peach X-disease phytoplasma, may be transmitted by several species of leafhoppers; others, such as elm yellows phytoplasma, appear to be transmitted by one or only a few species (Lee and Davis, 1992).

1.5. Phylogenetic position of phytoplasmas

Phytoplasmas have diverged from gram-positive eubacteria, and belong to the Genus phytoplasma within the Class Mollicutes and order Acholeplasmatales (Figure.1.5.). Currently the phytoplasma is at candidatus status which is used for bacteria that cannot be cultured.

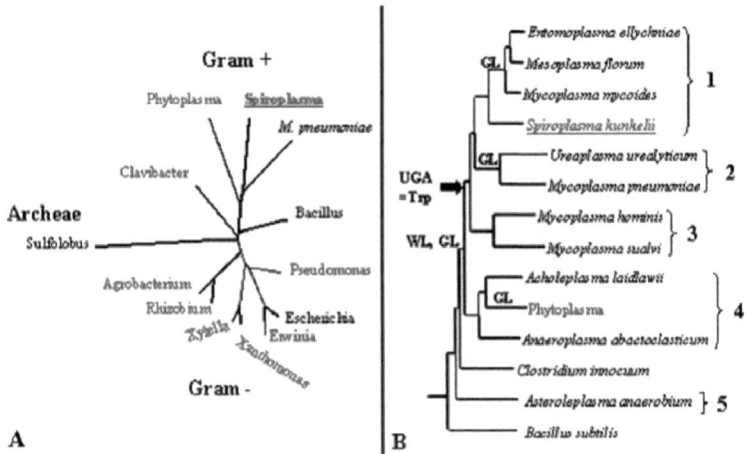

Figure.1.5. Phytoplasmas are firmicutes. A. Phylogenetic relationships of several bacterial clades containing bacterial pathogens. **B.** The 5 phylogenetic groups within the Class *Mollicutes*. Plant pathogenic/symbiotic bacteria are indicated in green. GL, gene loss; WL, loss of cell wall. Figure taken from Saskias phytoplasma website, http://www.jic.ac.uk/staff/saskia-hogenhout/index.htm.

Introduction

Recently, phylogenetic analyses based on 16S rRNA and ribosomal protein gene sequences have revealed that the uncultured phytoplasmas form a large discrete monophyletic clade within the class mollicutes(Gundersen et al., 1994). Phytoplasma taxonomic groups are based on differences in the fragment sizes produced by the restriction digest of the 16SrRNA gene sequence (RFLP) or by comparison of DNA sequences from 16S/23S spacer regions (Hodgettset al., 2007).

1.6. Management and control of phytoplasma diseases

Methods of control vary considerably from one disease to another, depending on the kind of pathogen, the host, the interaction of the two, and many other variables (Agrios, 2004).

Most serious diseases of crop plants appear on a few plants in an area year after year, spread rapidly, and are difficult to cure after they have begun to develop. Therefore, almost all control methods are aimed at protecting plants from becoming diseased rather than at curing them after they have become diseased (Agrios, 2004).

In controlling phytoplasmal diseases, the primary concern is often prevention rather than treatment due there is no known cure for phytoplasmal infections. However, infected plants or dormant propagative tissue can be freed of phytoplasma by heat treatment.

1.6.1. Prevention strategy

Propagate from seed or from phytoplasma-free plants, that means select propagating material from sources known to be free of disease or indexed free of disease.
Eliminate perennial and biennial weed hosts and eradicate known diseased trees as soon as they occur. Therefore, removal of phytoplasma infected plants eliminates sources of infection. Therefore, early, fast, specific and sensitive detection and diagnosis of phytoplasmas are very important for effective prevention strategies, especially because phytoplasmas may have a very long latency period. However, the most promising strategy for avoiding phytoplasma disease is the identification or development of resistant plant varieties (Welliver, 1999). In order to advance this field of research basic knowledge about the epidemiology, the pathogenicity mechanisms of the phytoplasmas, the effects of environmental factors on disease and symptom development, and the nature of resistance/tolerance in host plants is required

1.6.2. Control of insect vectors

"When the pathogen is introduced or spread by an insect vector, control of the vector is as important as and sometimes easier than, the control of the pathogen itself. In the case of viruses, phytoplasmas, and fastidious bacteria, however, of which insects are the most important spreading agents, insect control has been helpful in controlling the spread of their diseases only when it has been carried out in the area and on the plants on which the insects overwinter or feed before they enter the crop. Controlling such diseases by killing the insect vectors with insecticides after they have arrived at the crop has rare proved sufficient. Therefore, in cases where the insect vector is known and the time of its occurrence established, insecticide programs may be of value when directed at the vector before it becomes established on the plants. Typically, insecticide sprays are of limited value since migrating vectors may transmit the phytoplasma before the insecticide kills those" (Agrios, 2004).

1.6.3. Management strategy

Because phytoplasmas lack a cell wall , they are resistant against antibiotics that interact with cell wall synthesis like penicillin but other antibiotics with an alternative modes of action like tetracyclines can inhibit their growth (bacteriostatic to phytoplasmas). Therefore, remission of the disease symptoms can be achieved experimentally by injecting the antibiotic tetracycline but without continuous use of this antibiotic, disease symptoms will reappear again(Davies et al., 1968). In addition, antibiotic treatment is expensive and time-consuming. Therefore, the best strategy is to apply an efficient elimination program.

As a conclusion, the only dynamic way to control phytoplasma infection has been to prevent the emergence by guaranty that clean planting material is used, or by quest to find and/or breed varieties of crop plants that are resistant or tolerant to the phytoplasma/insect vector.

1.7. Anatomy of phloem cells

Phloem cells conduct soluble organic material made during photosynthesis in leaves to rest of the plant. They are alive at maturity and tend to stain green (with the stain fast green). Phloem cells are usually located inside the xylem. The two most common cells in the phloem are the companion cells and sieve tube cells.

Introduction

The sieve-tube cells lack a nucleus, have very few vacuoles, but contain other organelles. The sieve tube is an elongated rank of individual cells, called sieve-tube members, arranged end to end. The endoplasmic reticulum is concentrated at the lateral walls. Sieve-tube members are joined end to end to form a tube that conducts soluble organic food (photosynthates) materials throughout the plant. The end walls of these cells have many small pores and are called sieve plates and have enlarged plasmodesmata (Esau, 1965).Companion cells retain their nucleus and control the adjacent sieve cells (Figure.1.6).

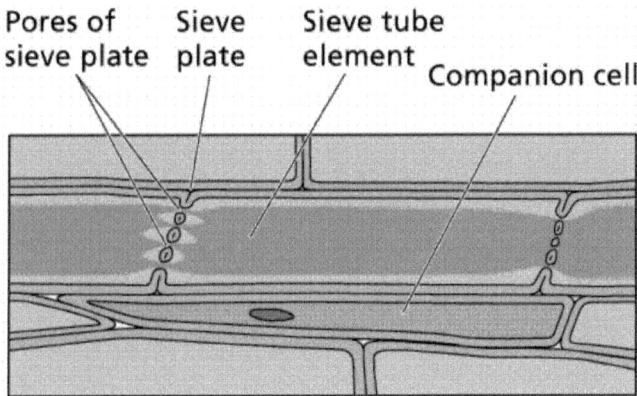

Figure.1.6. Diagram of the longitudinal view of phloem cells. This image is from Purves et al., (1992). Life: The Science of Biology, 4th Edition.

1.8. Some phytoplasma diseases of sugarcane

1.8.1. Sugarcane yellow leaf syndrome

Yellows diseases have been known since the early 1900s. One such disease, aster yellows, was first reported in 1902. Before 1967, its causal agent was thought by plant pathologists to be of viral origin because it could not be cultured in artificial media.

Sugarcane yellow leaf syndrome (YLS), characterized by a yellowing of the midrib and lamina, (Figure.1.7), was first reported in the 1960s from East Africa (Rogers, 1969) and later from Hawaii (Schenck, 1990), South Africa (Cronje et al., 1998) and Cuba (Arocha et al., 1999). It is now widely distributed in most sugarcane growing countries from all continents. Losses from 30% to over 60% of susceptible varieties have been reported (Schenck et al., 1997; Comstock et al., 1994; Arocha et al., 2000). Symptoms of YLS have been attributed to many causes, both biotic and a biotic, but the biotic causes are associated with infection by luteovirus or by phytoplasmas in Hawaii, Brazil, Australia, South Africa, Cuba, the USA and Mauritius.

Phytoplasmas have been consistently associated with YLS, but latent infections also occur (Bailey et al., 1996; Cronje et al., 1998; Arocha, 2000; Aljanabi et al., 2001).

Figure.1.7. Sugarcane yellow leaf syndrome (YLS). Sugarcane yellow leaf syndrome is characterized by a yellowing of the midrib and lamina. Symptoms consist of yellowing leaves with a bright yellow midrib, often when the rest of the lamina is still green. This picture was taken from Komor et al., 2010.

1.8.2. Sugarcane white leaf and sugarcane grassy shoot

Sugarcane white leaf (SCWL) and sugarcane grassy shoot (SCGS) occur only in the south-east Asian region and not in the other sugarcane growing areas of the world. Both are caused by a single phytoplasma type that is a member of the SCWL group and appears to infect only sugarcane. The most characteristic symptoms of SCWL are the presence of leaves with total chlorosis, proliferating tillers and pronounced stunting. The leaves are narrower and smaller than those of healthy plants (Figure.1.8). SCWL is naturally transmitted by the leafhopper *Matsumuratettix hiroglyphicus* organism Matsumura (Matsumoto et al., 1968). Records on mechanically transmission as well as on transmission by aphids have not been confirmed (Rishi and Chen, 1989).

Introduction

Sugarcane grassy shoot (SCGS) is one of the most important diseases of sugarcane in India. It was first observed in 1949 (Chona, 1958).SCGS has been recorded in most sugarcane-growing areas of India and is known to occur also in Thailand (Wongkaew et al., 1997; Sdoodee et al., 1999). SCGS disease is characterized by the production of a large number of thin, slender, adventitious tillers from the base of the affected stools. This profuse growth gives rise to a dense or crowded bunch of tillers bearing pale yellow or chlorotic leaves which remain thin, narrow, reduced in size and have a soft texture. The vector(s) responsible for the natural spread of SCGS have not been identified. There are reports on transmission of SCGS by three different species of aphids as well as by the fulgorid Proutista moesta Westwoo(Chona et al., 1960; Edison et al., 1976). However, these reports have not been confirmed (Rishi and Chen, 1989).

Figure.1.8. Sugarcane white leaf (SCWL). SCWL disease is caused by phytoplasma in Thailand. The most characteristic symptoms of SCWL are the presence of leaves with total chlorosis, proliferating tillers and pronounced stunting. The leaves are narrower and smaller than those of healthy plants. This picture was taken from Komor et al., 2010.

1.9. Detection of sugarcane phytoplasma infections

Sugarcane phytoplasma infections can be detected by microscopic examination of phloem tissue sections stained with the DNA fluorochrome4-6 diamidino-2-phenylindole (DAPI) (Seemüller, 1976; Sarindu and Clark, 1993). This procedure is simple, rapid and not much expensive. However it is limited when the phytoplasma population is very low and unevenly distributed among the plant host organs, as is often true for sugarcane.

For detection and identification of sugarcane phytoplasmas, the powerful PCR technology has widely been employed in several laboratories. It offers several advantages over other methods including versatility, relative simplicity, specificity and high sensitivity. Primers amplifying rRNA gene sequences proved most suitable for PCR. It may be performed as one-round PCR or by reamplifying the DNA fragments obtained in the first amplification using internal primers (nested-PCR). Very often in affected sugarcanes the phytoplasma numbers are so low that infections could be identified only through the highly sensitive nested PCR assay (Tran-Nguyen et al., 2000; Aljanabi et al., 2001).

1.10. Sugarcane yellow leaf syndrome in Hawaii

A novel sugarcane disease was observed in Hawaiian sugarcane plantations in the 1990s, characterized by a yellowing of the leaf midrib, which was followed by stunted leaf tops and yield decline (Schenck, 1990, Lehrer et al., 2009). Similar symptoms were reported shortly later from plantations in Brazil, mainland USA and South Africa (Vega et al., 1997; Comstock et al., 1994; Bailey et al., 1996).

The disease was called Yellow leaf syndrome (YLS) and classified in 2000 as a "disease of unknown origin" (Lockhart and Cronje, 2000). Research on the nature of the causal agent was controversial among plant pathologists. A RNA-virus was isolated from symptomatic plants and named *Sugarcane yellow leaf virus* (SCYLV). It was proposed as the causal agent for YLS (Borth et al., 1994; Vega et al., 1997). A survey of YLS-diseased sugarcane plants in Africa failed to reveal a close correlation between SCYLV and symptoms, a better correlation was seen between the presence of a phytoplasma infection and symptom expression (Cronje et al., 1998). The phytoplasma-caused disease was called Leaf yellows (LY) in contrast to the virus-caused disease, which is now called Yellow leaf (YL).

The phytoplasma was named *Sugarcane yellow leaf phytoplasma* (SCYLP). It was found in sugarcane from Australia, South Africa, Cuba, India and Mauritius (Arocha et al., 1999, 2005a; Cronje et al., 1998; Aljanabi et al., 2001; Gaur et al., 2008), sometimes together with SCYLV. Severe sugarcane diseases in South-East Asia and Africa are known to be caused by phytoplasma (Marcone, 2002), for example White leaf (Chen and Kusalwong, 2000), Grassy shoot (Viswanathan, 2000), Green grassy shoot (Pliansinchai and Prammanee, 2000) and

Ramu stunt (Suma and Jones, 2000). Twenty-five different phytoplasma isolates were obtained from North Australian sugarcane plants and none of them was closely related to White leaf and Grassy shoot, although also none of them could be related to sugarcane disease symptoms (Tran-Nguyen et al., 2000).

Many publications deal with the *Sugarcane yellow leaf virus* and the associated disease, for example its worldwide distribution (Abu Ahmad et al., 2006; Komor et al., 2010), its nucleotide sequence (Moonan et al., 2000; Smith et al., 2000), the transmission to the plant (Schenck and Lehrer, 2000; Lehrer et al., 2007) and the physiological effects on the infected plant (Yan et al., 2009). The virus-caused leaf yellowing syndrome is now accepted as an important, worldwide threat for sugar yield (Grisham et al., 2002; Lehrer et al., 2009). Also the South African sugar industry, for which originally the phytoplasmas were thought to be the main reason for YLS, appears to be predominantly infected by SCYLV and not by phytoplasma (Rutherford et al., 2004). However, the YLS-problem is not fully solved yet.

The Hawaiian sugarcane cultivars were differentiated into so-called susceptible cultivars which contain relatively high titres of SCYLV, and resistant cultivars with 100 time's lower virus titres (Zhu et al., 2010). Experiments with infected and virus-free plants of the same cultivar indicated that the viral infection led to higher symptom expression and to yield losses (Lehrer and Komor, 2008). The picture became less clear, when susceptible cultivars (i. e. with high virus-titre) and resistant cultivars (i. e. with low virus titre) were compared. The correlation between symptom expression and SCYLV-presence was not strict, some strongly infected cultivars exhibited relatively little symptoms and some resistant cultivars showed symptoms, although at low intensity (Lehrer and Komor, 2008). Therefore the question arose, whether some of the Hawaiian sugarcane cultivars were also infected by *Sugarcane yellow leaf phytoplasma* (SCYLP), thus causing leaf yellowing symptoms independent of or together with SCYLV. The simultaneous presence of SCYLV and SCYLP was reported to aggravate the leaf yellowing symptom expression in sugarcane (Aljanabi et al., 2001).

So far sugarcane white leaf or sugarcane grassy shoot symptoms had not been reported in Hawaiian plantations; however the presence of a low-symptom pathogen such as SCYLP may have escaped attention of breeders and growers. There are two reports about phytoplasma diseases in Hawaii, one about water cress yellows caused by an Aster yellows type phytoplasma and transmitted by an accidentally introduced leaf hopper (Borth et al., 2002; 2006), the other about a yellows disease of a native tree, *Dodonaea viscosa*, caused by a Western X-disease phytoplasma (Borth et al., 1995).

We tested Hawaiian cultivars (and for comparison a few cultivars from Cuba, Egypt and Syria) for phytoplasma to reveal whether sugarcane phytoplasma is around in Hawaii and in Hawaiian sugarcane plantations.

The main objectives of this project were to determine the following:

1. Possible association of phytoplasma(s) infection with YLS symptoms in sugarcane plants, using of molecular techniques, namely PCR, for a more accurate determination.

2. Which type(s) of phytoplasma(s) are associated with YLS symptoms in sugarcane plants from Hawaii breeding station, Hawaii plantations, Cuba, Middle East and areas in Thailand?

3. How does this type(s) compare to other known phytoplasma types by phylogenetic analysis?

4. Could it be that Hawaiian plantations have phytoplasma after hot water treatment?

5. Can sugarcane aphid (*Melanaphis sacchari*) transmit the detected phytoplasma to sugarcane plants?

6. Ultrastructural changes of the leave anatomy and cytology by phytoplasma infection and cytological location of phytoplasma.

2. Material and Methods

2.1. Material

Balance (Mettler P1210)

Centrifuges, Type centrifuge 5403 (Eppendorf)

Centrifuge, Type Mikro 20 (Hettich)

Centrifuge, Type UNIVERSAL 32R (Hettich)

Diamond knife (type 35°, Diatome, Biel, Switzerland)

Electron microscope, Type ZEISS 902 (Zeiss, Oberkochem)

Gel Electrophoresis, Type GNA 100 (Pharmacia LKB)

Gene power supply, type GPS 200/400 (Pharmacia)

Mini-Vertical Gel Electrophoresis, Type SE 250 and SE 260 (Mighty small II)

Thermomixer comfort (Eppendorf)

Thermal cycler, Type PTC- 100 (MJ Research)

Thermal cycler, Type Master cycler personal, with heated lid and 1 personal card, 115 V/60 Hz (Eppendorf)

Thermal cycler, type MyiQ qPCR detection system for single-colour experimentation (Bio-rad)

Spectrophotometer, Type 650 (Beckman)

Ultra cut microtome (Leica Microsystems, Wetzlar, Germany)

Vortexer, Type REAX-1R (Heidolph)

Heated magnetic stirrer, Type MR 82 (Heidolph)

Microwave oven, Type KOR- 6115 (Alaska)

Nanophotometer, Type UV/Vis spectrophotometer (Implen)

pH-mV-meter , Type 531 (Knick)

UV-SYSTEME (NTAS)

2.1.2. Chemicals and Enzymes

2.1.2.1. Chemicals

Agarose NEOO (Carl Roth GmbH)

30% Acrylamide

10% Ammonium Persulfate

BSA (Sigma-Aldrich Chemie Gmbh)

Chloroform-isaomyl alcohol

dNTP Set, molecular biology grade (MBI Fermentas)

Ethanol

Phenol

Polyvinyl pyrrolidone (PVP) (Sigma-Aldrich Chemie Gmbh)

2.1.2.2. Enzymes

Proteinase K (Roch Diagnostics GmbH)

RNase A (Promega GmbH)

RasI (MBI Fermentas)

HpaII (MBI Fermentas)

HinfI (MBI Fermentas)

KpnI (MBI Fermentas)

MesI (MBI Fermentas)

Taq DAN Polymerase (MBIFermentas)

2.1.3. Buffers, Solutions

2.1.3.1. Buffer and solutions for DNA extraction

Table.2.1. Ingredients of CTAB extraction buffer.

Final concentration	Reagents	For 100 mL	Stock
2%	CTAB	2 g	
2%	PVP(MW 40000)	2 g	
1.4 M	NaCl	28 ml	5 M
20 mM	EDTA pH 8.0	4 ml	0.5 M, pH 8.0
100 mM	Tris HCl pH 8.0	10 ml	1 M, pH 8.0
2%	2-mercaptoethanol	2 ml	

The solution was prepared without CTAB, PVP and ß-mercaptoethanol and autoclaved for 20 min. When needed add 2% CTAB (w/v) , 2% PVP (w/v) and 2% ß-mercaptoethanol and heated at 65°C in order to dissolve CTAB and PVP.

Phenol: chloroform: isoamyl alcohol (25:24:1)

Chloroform: Isoamyl alcohol (24:1 v/v)

Isopropanol 100%

Material and Methods

5M NaCl 292.2 g/L
RNase A (Promega GmbH)
Ethanol 100%
Ethanol 70%
TE 1X (10 mM Tris, 1 mM EDTA, pH 8.0)

2.1.3.2. Buffer for gel electrophoresis

TBE 10X (108 g Tris base, 55 g Boric acid, 40 ml 0.5 EDTA, pH 8.0, H_2O was added to final volume 1Liter.)

2.1.3.3. Buffer for PCR

10X Taq Buffer with KCl and 15mM $MgCl_2$ (100mM Tris-HCl, pH 8.8, 500mM KCl, 0.8% (v/v) Nonidet P40 and 15 mM $MgCl_2$)

10X Taq Buffer with (NH4)2So4 and 20mM $MgCl_2$ (750mM Tris-HCl, pH 8.8, 200mM (NH4)2SO4 and 0.1% (v/v) Tween 20 and 20mM $MgCl_2$

$MgCl_2$ 1M (203g $MgCl_2.6H_2O$, add H_2O to 1L)

2.1.3.4. Buffer for restriction enzymes

10X Buffer Tango (33 mM Tris-acetate, pH 7.9 at 37°C, 10 mM Mg-acetate,
66 mM K-acetate and 0.1 mg/ml BSA)

10X Buffer KpnI (10 mM Tris-HCl, pH 7.5 at 25 c, 10 mM $MgCl_2$, 0.02% Triton X-100 and 0.1 mg/ml BSA)

2.1.3.5. Buffer for polyacrylamide gel electrophoresis

Acrylamide: bisacrylamide (29:1) (% w/v) (29g acrylamide, 1g N,N-methylenebisacrylamide, H_2O to 100 ml)

1X TBE electrophoresis buffer (89 mM Tris-borate, 2mM EDTA, pH 8.0)

TBE is usually made and stored as a 5% stock solution. The pH of the Buffer should be approximately (8.3). Polyacrylamide gels are poured and run in 0.5x or 1xTBE at low voltage (1-8 V/cm) to prevent denaturation of small fragments of DNA by Joulic heating. Other electrophoresis buffers such as 1x TBE can be used, but they are not as good as TBE.

10 %(w/v) Ammoniumpersulfate (ammonium persulfate 1g, H_2O to 10 ml).This solution is used as a catalyst for the copolymerization of acrylamide and bisacrylamide gels.

This solution may be stored at 4°C for several weeks.

TEMED: electrophoresis grade TEMED is available from Bio-Rad, Sigma, and other suppliers. This solution is stored at 4°C.

2.1.4. Kits

2.1.4.1. Isolation of Nucleic Acids for PCR

High Pure PCR Template Preparation Kit (Roche Diagnostics GmbH)
Genomic DNA purification Kit (MBI Fermentas)

2.1.4.2. Nucleic acids purification

Agarose Gel DNA Extraction Kit (Roche Diagnostics GmbH)
High Pure PCR product purification Kit (Roch Diagnostics GmbH)

2.1.4.3. Q-PCR

SensiMix II probe (2X) kit (BIOLINE)
SsoFast ProbesSupermix kit (BIO-RAD)

Material and Methods

2.1.5. Oligonucleotides

Table.2.2. Sequences of universal primers used in the amplification of phytoplasma 16S rRNA operon.

Primer	Sequence	Reference
P1 (Forward)	5'-AAGAGTTTGATCCTGGCTCAGGATT-'3	Deng and Hiruki (1991)
P7(Reverse)	5'-CGTCCTTCATCGGCTCTT-'3	Smart et al. (1996)
P4(Forward)	5'-GAAGTCTGCAACTCGACTTC-'3	Smart et al. (1996)
R16F2n(Forward)	5'- GAAACGACTGCTAAGACTGG-'3	Lee et al. (1993)
R16R2(Reverse)	5'- TGACGGGCGGTGTGTACAAACCCCG-'3	Lee et al. (1993)
U-1 (Forward) SN910601(Forward)	5'-GTTTGGATCCTGGCTCAGGATT-'3	(Namba et al. 1993; Wongkaew et al. 1997)
MLO-7 (Reverse)	5'-CGTCCTTCATCGGCTCTT-'3	,,
MLO-X (Forward)	5'-GTTAGGTTAAGTCCTAAAACGAGC-'3	"
MLO-Y (Reverse)	5'-GTGCCAAGGCATCCACTGTATGCC-'3	"
P1 (Forward)	5'-GTCGTAACAAGGTATCCCTACCGG-'3	"
P2 (Reverse)	5'- GGTGGGCCTAAATGGACTTGAACC-'3	"
SN910601(Forward)	5'-GTTTGATCCTGGCTCAGGATT-'3	"
P6 (Reverse)	5'-CGGTAGGGATACCTTGTTACGACTTA-'3	(Deng & Hiruki, 1991)

2.1.6. Software for Gene analysis

The obtained nucleotide sequences were compared with sequences of phytoplasmas and acholeplasmas from GenBank using the Blastn program (http://blast.ncbi.nlm.nih.gov).

BioEdit (http://www.mbio.ncsu.edu/BioEdit/bioedit.html) and MUSCLE software version 3.8 (Edgar, 2004) were used for sequence comparison and alignment (http://ebi.ac.uk/tools/mas).

The phylogenetic trees were constructed by maximum liklihood analysis with geneious program (http://www.geneious.com) through the PhyML software (http://atqc.lirmm.fr /phyml/) (Guindon and Gascuel, 2003).

2.2. Methods

2.2.1. Plant material

Sugarcane is a member of the family Gramineae and it belongs to the genus *Saccharum* (S.).*Saccharum Officinarum* was used throughout the present investigations. Cultivars were obtained as stem cuttings from different sources and were grown in pots in the greenhouse at temperatures between 22°C during night and up to 27°C in the sunny days under greenhouse conditions. The dry plants were watered with tap water every day. Sugarcane was propagated by planting the single-bud (cutting) pieces in sterilized vermiculite and were placed in the climate chamber at 28°C with very high humidity conditions for about 15 days .Next, produced plants (seedling plants) were transferred outside the climate chamber into small plastic pots each holding soil composed of bark humus, plant humus, peat, pumice stone, expanded clay, one plant per pot. When the plants were 70-day old, they were transferred to bigger pots, each holding clay soil/vermiculite (1:1). In order to assess the relationship between expression of YLS symptoms and the presence of phytoplasmas, leaves were collected from all sugarcane varieties grown in pots in greenhouse.

2.2.2. DNA extraction strategies

The extraction of DNA from the samples is necessary for the molecular analyses that follow. Total nucleic acid, for use as templates in PCR, was extracted from fresh tissue or from frozen tissue according to the methods described by Harrison et al., (1994) and also Doyle and Doyle (1990) with some modifications. Like many other plant species, sugarcane tissues contain high levels of polysaccharides and polyphenolic compounds, which present a major contamination problem in the purification of plant DNA. When cells are disrupted these cytoplasmic compounds can come into contact with nuclei and other organelles (Loomis, 1974).

(1-2) grams of frozen tissues or fresh tissues were cut into small pieces and ground to a fine powder in liquid nitrogen using a pre-chilled mortar and pestle.
The powdered tissues (up 200-300 mg) were transferred to 1.5 or 2 ml eppendorf microfuge tubes then 500 µl CTAB extraction buffer pre-warmed at 65°C to each sample was added .
The tissues were suspended (wetted) by gentle shaking.

The samples were incubated at 65°C for 60 min or more in a water bath or in an incubator with occasional gentle mixing with inversion 3-4 times during incubation to dissolve all nucleic acids, then allowed cooling for a few min.

The extracts were then mixed with an equal volume phenol, chloroform, isoamyl alcohol (25:24:1) and centrifuged at 10.000 rpm for 10 min.

The supernatants were transferred to new tubes and an equal volume (500 µl) of chloroform-isoamyl alcohol was added to each sample and gently mixed for 5 min to form emulation.

The samples were centrifuged at 14000 rpm for 10 min (long enough to produce a clear supernatant).

The upper aqueous layers (containing the DNA) were carefully transferred to new tubes (with avoiding taking up any of the interface material).

(Optional) RNase step: RNase was added to the aqueous contents of each tube and incubates at 37°C for 30 min.

The samples were re-extracted for second time by using slightly less chloroform-isoamyl alcohol (250 µl) (half as much as first time).

The volumes of extracts were estimated then 2 volume of 100% cold ethanol and one tenth volume of 3M sodium acetate (pH = 5.2) or 5M NaCl were added to each sample and mixed gently by inverting.

Total nucleic acids were precipitated after incubation at -20°C for 1hr to overnight.

The samples were centrifuged at 10000 rpm for 10 min then the alcohol supernatant carefully discarded.

The pellets were washed twice with 70% ethanol, air dried and resuspended in sterile distilled water or TE buffer.

DNA solutions were then stored at -20°C until use.

After DNA samples are dissolved, 2µl of each sample were checked on 1% agarose gel in order to evaluate template integrity. Next the DNA concentrations were measured spectrophotometrically.

The total nucleic acid was also extracted by using High Pure PCR Template preparation Kit (Roche) or by using Genomic DNA purification Kit (MBI Fermentas).

2.2.3. Polymerase Chain Reaction (PCR) for the detection of phytoplasmas

Symptomatology had been one the major criteria for diagnosing the phytoplasma disease before molecular-based methods become available. It remains the important clue used for preliminary identification of putative phytoplasmal diseases.

The polymerase chain reaction (PCR) is a rapid procedure for in vitro enzymatic amplification of a specific segment of DNA (Donaldet al., 2006). PCR has been used during the last years for the detection of large number of microorganisms, also including phytoplasmas. Several universal primer pairs designed for the amplification of the 16SrRNA gene of phytoplasmas were tested. The method found to give consistent results was the nested PCR (Heinrich et al., 2001; Srivastava et al., 2005).

2.2.3.1. Definition of Nested PCR

Nested PCR is a variation of the polymerase chain reaction (PCR), in that two pairs (instead of one pair) of PCR primers are used to amplify a fragment. The first pair of PCR primers amplifies a fragment similar to a standard PCR. However, a second pair of primers called nested primers (as they lie) are nested within the first fragment) bind inside the first PCR product fragment to allow amplification of a second PCR product which is shorter than the first one (Pérez de Rozas et al., 2008), (Figure.2.1).

Material and Methods

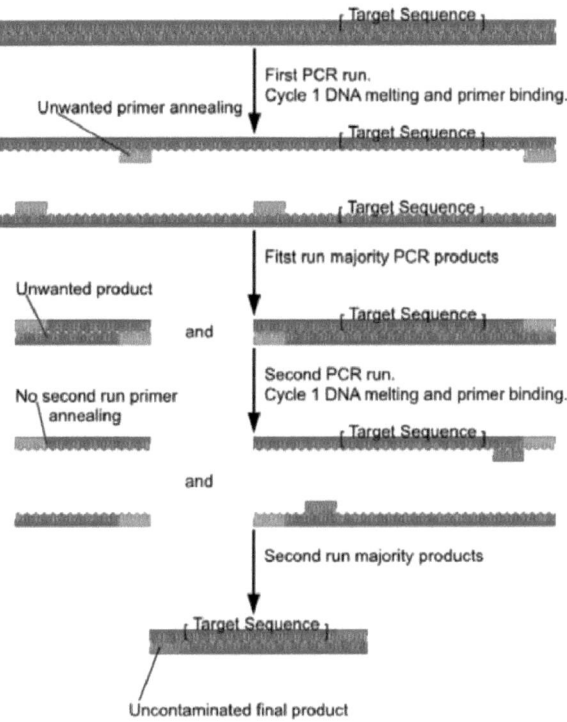

Figure.2.1. A diagram illustrating of the method of nested PCR. Figure taken From Wikipedia, the free encyclopaedia.

The advantage of nested PCR is that if the wrong PCR fragment was amplified, the probability is quite low that the region would be amplified a second time by the second set of primers. Thus, nested PCR is a very specific PCR amplification. Furthermore, the double amplification in the nested-PCR increases the sensitivity of PCR reaction in 2-3 logarithmic units when compared with conventional PCR (Lindqvist, 1999; Marsilio et al., 2005).

2.2.3.2. Nested PCR Reaction

Nested PCR requires two sets of primers which are used to amplify a specific DNA fragment using two separate runs of PCR. A standard reaction mixture of 25 µl consisted of the following:

10X Taq buffer

dNTP mix (200 µM each dNTP)

Forward and reverse primers (0.4 µM)

Taq DNA polymerase (5 U/µl)

Template DNA (100 ng)

dd H_2O to final volume 25 µl

2.2.3.3. First round of PCR

Nucleic acid samples were diluted in sterile distilled water to give a final concentration of 100 ng/ul and in some cases DNA concentrations were not adjusted after extraction, but used as isolated, 1 µl of DNA solution was used per reaction tube.

2.2.3.4. Nested round of PCR

One micro litre of diluted (1:30 or 1:20) PCR products from the first round was used as the template in the second amplification. In most cases first PCR products were used with any dilution. The PCRs (30 cycles) were done with an automatic thermal cycler in 25µl reaction tubes. Several universal primer pairs, which were previously designed, based on the phytoplasma rRNA operon, for the amplification of phytoplasmal DNA were tested (Figure.2.2 andFigure.2.3). The method found to give consistent results was the nested PCR. However, it was amazing when some primer pairs didn't work continually and we had to test in this case other primer pairs to check if the negative results were false due the primer pairs or due the phytoplasmas disappeared from our greenhouse sugarcane plants. Since this phenomenon may occur especially in our greenhouse when there are no insect vectors for phytoplasmas and the plants reproduce by vegetative propagation therefore, the titre of phytoplasma would be lower generation by generation.

2.2.3.5. Nested-PCR assay (I)

The primer pair combination used in the first round was P1/P7 while the nested primer pair was R16F2n/ R16R2 (Table.2.3).

Parameters of the PCR assays using external primer pair (P1/P7) were: denaturation step at 94°C for 30 s (4 min for the first cycle), annealing for 1.5 min at 55°C and primer extension for 1.5 min (10 min in final cycle) at 72°C.

Parameters of the PCR assays using internal(nested) primer pair (R16F2n/R16R2) were: denaturation step at 94°C for 30 S (4 min for the first cycle), annealing for 1.5 min at 56°C and primer extension for 1.5 min (10 min in final cycle) at 72°C.

Table.2.3. Oligonucleotide primers used for nested-PCR assay I.

Primer	Location	Type of PCR
P1 (Forward)	16S	First
P7 (Reverse)	23S	First
R16F2n (Forward)	16S	Nested
R16R2 (Reverse)	16S	Nested

2.2.3.6. Nested-PCR assay (II)

The primer pairs combinations used in the first and nested rounds of nested-PCR assay (II) are indicated in (Table.2.4).

Parameters of the PCR assays using external primer pair SN910601/P6 were: denaturation step at 94°C for 30 s (4 min for the first cycle), annealing for 1 min at 54°C and primer extension for 1.5 min (10 min in final cycle) at 72°C.

Parameters for PCR using internal primer pair R16F2n/R16R2, which amplifies 1250bp DNA fragment, were: denaturation step at 94°C for 30 s (4 min for the first cycle), annealing for 1 min at 56°C, and primer extension for 1.5 min (10 min in final cycle) at 72°C.

Table.2.4. Oligonucleotide primers used for nested-PCR assay II.

Primer	Location	Type of PCR
SN910601 (Forward)	16S	First
P6 (Reverse)	16S	First
R16F2n (Forward)	16S	Nested
R16R2 (Reverse)	16S	Nested

2.2.3.7. Nested-PCR assay (III)

The primer pairs combinations used in the first and nested rounds of nested-PCR assay (III) are indicated in (Table.2.5).

Material and Methods

The parameters of the PCR assays using external primer pair MLO-X/MLO-Y were: denaturation step at 94°C for 30 s (4 min for the first cycle), annealing for 1 min at 58°C and primer extension for 1.5 min (10 min in final cycle) at 72°C.

The parameters for PCR using internal primer pair P1/P2, which amplifies 210 bp DNA fragment, were: denaturation step at 94°C for 30 s (4 min for the first cycle), annealing for 45 s at 62°C, and primer extension for 1 min (10 min in final cycle) at 72°C.

Table.2.5. Oligonucleotide primers used for PCR assay III.

Primer	Location	Type of PCR
MLO-X (Forward)	16S	First
MLO-Y (Reverse)	spacer region (near 23S)	First
P1 (Forward)	16S (near the spacer region)	Nested
P2 (Reverse)	"tRNA-Ile" (near the spacer region)	Nested

2.2.3.8. Nested-PCR assay (IV)

The primer pairs combinations used in the first and nested rounds of nested-PCR assay (IV) are indicated in (Table.2.6).

Parameters of the PCR assays using external primer pair U-1/ MLO-7 were: denaturation step at 94°C for 30 s (4 min for the first cycle), annealing for 1 min at 56°C and primer extension for 1.5 min (10 min in final cycle) at 72°C.

Parameters for PCR using internal primer pair MLO-X/MLO-Y, which amplifies 700bp DNA fragment, were: denaturation step at 94°C for 30 s (4 min for the first cycle), annealing for 1 min at 60°C, and primer extension for 1.5 min (10 min in final cycle) at 72°C.

Table.2.6. Oligonucleotide primers used for PCR assay IV.

Primer	Location	Type of PCR
U-1 (Forward)	16S	First
MLO-7 (Reverse)	23S	First
MLO-X (Forward)	16S	Nested
MLO-Y (Reverse)	spacer region (near 23S)	Nested

Material and Methods

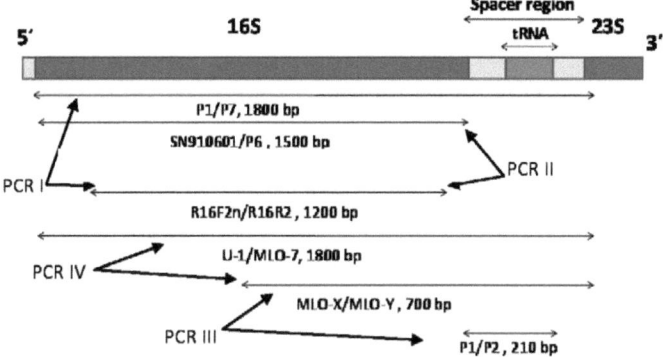

Figure.2.2. Diagrammatic representation of location of used primer pairs and expected size of their amplified products based on phytoplasma rRNA operon.

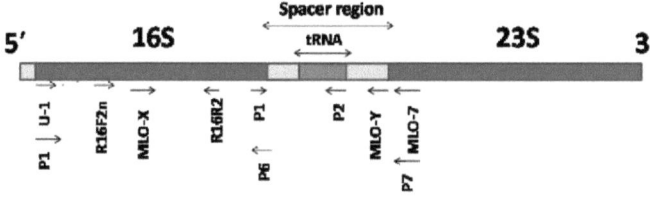

Figure.2.3. Diagrammatic representation of a phytoplasma rRNA operon and genomic location of primers used for phytoplasma detection.

2.2.4. Agarose Gel Electrophoresis

PCR products were electrophoresed on 1% agarose gel, stained with ethidium bromide and DNA bands visualized using a UV transilluminator

Agarose gel electrophoresis was used to visualize and isolate DNA molecules following PCR amplification. Agarose (1%) was dissolved in TBE buffer by heating in a microwave.

After cooling, 1 µ of a 1 mg/ml ethidium bromide solution was added per 50 ml gel and the gel was poured. Gels were run at 80-100 V for 1 hour.

2.2.5. Digestion of nested-PCR products

By RFLP analysis of PCR-amplified 16S rRNA gene, the phytoplasmas detected can be differentiated and classified (Lee et al., 1993). The basic technique for detecting RFLPs involves fragmenting the samples of DNA or (PCR products) by the restriction enzymes. Restriction enzymes recognize specific nucleotide sequences and cleave DNA molecules at a position either within or outside their recognition site (Roberts and Kenneth, 1976). These enzymes are important tools for numerous applications, including restriction fragment length polymorphism (RFLP) analysis of PCR products .The resulting DNA fragments are then separated by their length through gel electrophoresis.

RFLP analysis of PCR-amplified 16S rRNA gene sequences with a number of restriction enzymes was used by Lee et al., (1993) and Schneider et al., (1993) to differentiate various phytoplasmas by their distinct RFLP patterns. This procedure proved to be simple, reliable, and practical.

Our nested PCR products were analyzed by single enzyme digestion with different restriction endonucleases HpaIII, HinfI, KpnI, MesI and RsaI (MBI Fermentas).

The reaction mixture (30 µl) consisted of the following:
Reagents:
10 µl PCR products
2µl 10X recommended buffer for restriction enzyme
1-2 µl (10-20 u) restriction enzymes
17 µl water nuclease free
The reaction mixtures were incubated in the incubator at 37°C for 3-16 h.

2.2.5.1. Inactivation of restriction enzymes

Inactivation of restriction enzymes following a digestion reaction is often required for downstream applications. Thermal inactivation is a convenient method used to terminate enzyme activity. The majority of restriction enzymes can be heat-inactivated at 65°C or 80°C in 20 min. Digested products were separated by electrophoresis on 5% polyacrylamide gels. Next RFLP patterns were compared with those previously published.

2.2.6. Polyacrylamide Gel Electrophoresis

Nondenaturing polyacrylamide gels are used for the separation and purification of fragments of double-stranded DNA.

2.2.6.1. Steps of operation

Assembling the apparatus and preparing the gel solution
Casting the gel
Loading the samples and running the gel

2.2.6.2. Special equipment

The SE 250 Mighty small II is a miniature vertical slab gel unit intended for rapid electrophoresis of nucleic acid samples of small volume.

2.2.6.3. Detection of DNA in polyacrylamide gels by staining

Unlike agarose gels, polyacrylamide gels cannot be cast in the presence of ethidium bromide because the dye inhibits polymerization of the acrylamide. However, ethidium bromide can be used to stain the polyacrylamide gel after electrophoresis. In order to detect of DNA the gels were gently submerged in the appropriate staining solution. We used just enough staining solution to cover the gel completely and the gels were stained for 30 min at room temperature. Then the gels were removed from the staining solution and placed on the UV transilluminator and photographed.

2.2.7. Sequencing and phylogenetic analysis of ribosomal DNA

By direct sequence analysis or RFLP analysis of PCR-amplified products, the phytoplasmas detected can be differentiated and classified. Several classification systems have been proposed either directly based on sequence analysis or indirectly, by RFLP analysis of PCR-amplified 16S rRNA gene.

In order to amplify the 16S/23S spacer region we used P1/P7 for first PCR and P4/P7 for second PCR (Smart et al., 1996). P4/P7 PCR product was purified from agarose gels using Agarose Gel DNA Extraction Kit (Roche Diagnostics GmbH). The DNA sample was sequenced in one direction using P4 primer. Unfortunately, this primer pair didn't work well for some cultivars. Therefore, we used MLO-X/MLO-Y for first PCR and P1/P2 for nested-PCR to amplify partially the 16S/23S spacer region. In order to amplify the 16SrRNA we used R16F2n/R16R2 for nested PCR. Nested-PCR products were cleaned up using High pure PCR products purification Kit (Roche). DNA samples were sequenced in both directions using nested primer pairs.

2.2.7.1. Sample Preparation for Value Read Service in Tubes

The value read is the service of choice for fast and reliable standard sequencing reactions. It is highly automated to allow rapid processing of plasmids or PCR products. We used 1.5 ml tubes (no additional sealing with Parafilm) for samples and primers and we used one tube per sequencing reaction. The DNA samples (purified PCR products) were dissolved in the elution buffer (10 mM Tris-HCl, pH 8.5) and the concentrations of these purified PCR products were adjusted to get the final concentration 10 ng/µl or 2 ng/µl in a minimum volume of 15 µl and the required primer concentrations were 2 pmol/µl with minimum total volume 15 µl. The DNA samples (purified PCR products) were direct-sequenced in an ABI 3730XL automated sequencer using the sequencing service of Eurofins MWG Operon Ebersberg, Germany (http://www.eurofinsdna.com/). Next, sequences were compared with others in GenBank database using BLAST program. The sequence data were deposited in GenBank.

2.2.8. Hot water treatment

"When a pathogen is excluded from the propagating material (seed, tubers, bulbs, nursery stock, grafts, and cuttings) of host, it is often possible to grow the host free of that pathogen for the rest of its life" (Agrios, 2004). Vegetative propagating material free of pathogens that are systemically distributed throughout the plants (viruses, viroids, and phytoplasmas) is obtained from mother plants that had been tested and shown to be free of the particular pathogen or pathogens. Furthermore, the new plants must be grown in pathogen- and vector-free soil and then be protected from airborne vectors.

Phytoplasma may be transmitted by the propagation of scions and or cutting collected from diseased plant. Therefore, in vegetatively propagated crops like sugarcane; phytoplasma can be readily spread to new locations through infected stem cutting if suitable precautions are not taken. These precautions include cold- and hot water treatment and tissue culture (Parmessur et al., 2002).

Material and Methods

A hot water treatment is an effective method for the control of number of plants pests and diseases (plant pathogens) including phytoplasmas.

Hot water treatment (HWT) has been proposed since 1966 by Caudwell at 30°C for 72 h in order to cure dormant woody plant material from phytoplasmas. Afterward other works showed the effectiveness of the treatment against these pathogens (Lherminieret al., 1990; Tassart-Subirats et al., 2003). However, (HWT) must be carefully applied because may interfere with the vitality of plant material. Thus, (HWT) of dormant canes or plants aims at phytoplasma elimination without any alteration in their vegetative development capacity. In addition the treatment demonstrates a positive effect of sanitation against several bacterial diseases, pests and insects (including eggs) which may be present on plant material.

2.2.8.1. Preparation of the plant material prior to the hot water treatment

Infected plants or dormant propagative organs can be totally freed of phytoplasmas by heat treatment. Infected plants are kept in growth chambers at 30°C to 37°C for several days, weeks, or months; but dormant organs are immersed in hot water (Agrios, 2004).

Soaking induces a thermal shock susceptible of modifying the physiological state of the plant material (breaking of bud dormancy, inducing storage losses). Therefore, in order to prevent a poor vegetative development, the plant material should be thermally prepared to the treatment by storage for 12 to 48 hours at room temperature in a humid and aerated chamber. Furthermore, the temperature after immersion and the treatment duration should be respected and after treatment, the plant material should be left to set back to room temperature (avoid direct contact with cold water).

2.2.9. Sugarcane aphid transmission test

2.2.9.1. Insect rearing

Melanaphis Sacchari (Sugarcane aphid) insects were provided from Hawaii Island. Colonies of *Melanaphis Sacchari* were established on phytoplasma-infected sugarcane plants.

2.2.9.2. Plant material

All test plants raised from single-eye setts that had received cold- and hot-water treatment or hot-water treatment were negative (phytoplasma free) when tested by nPCR prior to being used in transmission studies and all plants with typical symptoms of YLS that were used as source plants for the acquisition-access feeds were previously tested positive for phytoplasma.

2.2.9.3. Transmission tests

Phytoplasma-infected sugarcane plants were transferred to cage and the insects were given acquisition-access feeding on this fresh sugarcane leaves. After 45 days, (acquisition period and latency period), phytoplasma-free plant were transferred into the cage for inoculation feeding and these plants were kept for about 3 months after inoculation and then tested for phytoplasma infection. The transmission tests were repeated twice.

2.2.10. Q-PCR (real-time PCR) assay

Most universal as well as specific phytoplasma diagnostic protocols rely on nested PCR, which, although extremely sensitive, is also time-consuming and poses risks in terms of carry-over contamination between the two rounds of amplification (Weintraub and Jones, 2010).

Despite the development of protocols which overcome most the difficulties of phytoplasma diagnosis, the detection of these pathogens is still quite laborious. Q-PCR offers the opportunity to detect these pathogens in a sensitive and specific manner, bypassing all post-PCR manipulations. Therefore, direct qPCR has recently replaced the traditional PCR in efforts to increase the speed and sensitivity of detection and to improve techniques for mass screening (Weintraub and Jones, 2010). During a qPCR run, accumulation of newly generated amplicons is monitored at each cycle by fluorescent detection methods. The amount of fluorescence, monitored at each amplification cycle, is proportional to the log of concentration of the PCR target, and for this reason qPCR is also a powerful technique for the quantification of specific DNA.

2.2.10.1. Methods of monitoring DNA amplification in qPCR

The first method is fluorescent dyes (e.g. SYBR Green I).This double-stranded DNA binding dye emits a strong fluorescent signal when binding to double-stranded DNA. Therefore, during each subsequent PCR cycle more fluorescence signal will be detected(Figure.2.4). The main disadvantage of using a dye such as this is the lack of specificity. Therefore, wide optimization is required in this method. SYBR® Green I dye chemistry is not supported for plus/minus assays such as diagnosis of phytoplasmas.

Material and Methods

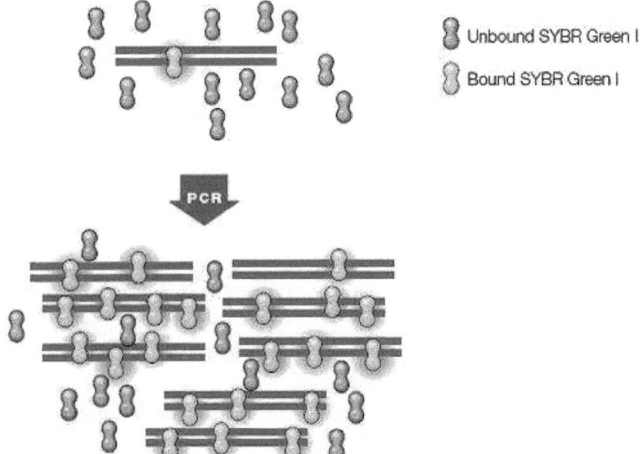

Figure.2.4.Diagram illustrating of SYBR Green during PCR amplification. Figure taken from BIO-RAD gene expression getaway.

The second method is fluorescent probes including hydrolysis (TaqMan) probes.

In this method, the hydrolysis probe is labelled with a quencher fluorochrome, which absorbs the fluorescence of the reporter fluorochrome as long as the probe is intact. However, upon amplification of the target sequence, the hydrolysis probe is displaced and subsequently hydrolyzed by the Taq polymerase. This results in the separation of the reporter and quencher fluorochrome and consequently the fluorescence of the reporter fluorochrome becomes detectable(Figure.2.5). During each consecutive PCR cycle this fluorescence will increase due of the progressive and exponential accumulation of free reporter fluorochromes.

Material and Methods

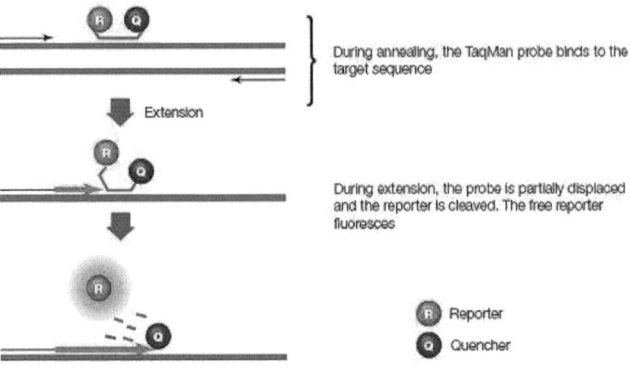

Figure.2.5.Diagram illustrating of TaqMan probe chemistry mechanism. Figure taken from BIO-RAD gene expression getaway.

The main advantages of using TaqMan probes include high specificity, a high signal-to-noise ratio, and the ability to perform multiplex reactions. Specific amplification of target sequences is directed by custom designed primers and probes. The first degree of specificity is achieved by the combination of amplification primer sequences. An additional degree of specificity results from a probe that hybridizes to a region of nucleic acid sequence that identifies the microbe of interest. Therefore, TaqMan probes are the most commonly used ones for the diagnosis of phytoplasmas.

2.2.10.2. Detection of phytoplasma based on TaqMan qPCR assays

Q-PCR assays were performed in optical 96-well plates with optical adhesive covers using a Bio-Rad iCycler thermal cycler (My iQ Optical Module) in a total volume of 50 µl or 20µl, including 5 µl or 2 µl respectively of DNA extracts (concentrations were 100 to 200 ng/µl) and TaqMan core reagents consisting of SensiMix II probe (2X) kit (BIOLINE) or SsoFast Probes Supermix kit (BIO-RAD). All primers were used at a final concentration of 400 nM and all probes at a final concentration of 100 nM. All primers and probes, which previously designed (Christensen et al., 2004)(Table 2.7 and Figure.2.6), were synthesized by Eurofins MWG /operon (Ebersberg, Germany), and all probes were lablelled at the 5´end with the fluorescent dye 6-carboxyfluorescein (FAM) as reporter and at 3´ end with 6-tetramethylrhodamine (TAMRA) as fluorescent quencher. Each sample was tested in triplicate and negative controls containing nuclease-free water in the place of DNA were included in all runs in order to test possible contamination. In additions, phytoplasma-infected

Material and Methods

periwinkle and phytoplasma-infected sugarcane KK34 were used as positive controls for both phytoplasma assay and plant assay. The thermal cycling conditions were 10 min at 95°C for one cycle as initial activation ; followed by 40 cycles each one consisting of two-step; 15 s for denaturation at 95°C and 1 min at 60°C for annealing and extension.

Table.2.7. Sequence of primers and probes used for detection of phytoplasma and plant DNA (Christensen et al., 2004).

Primer or probe	Phytoplasma 16S rRNA gene	Plant 18S rRNA gene
Forward primer	5'-CGTACGCAAGTATGAAACTTAAAGGA-'3	5'-GACTACGTCCCTGCCCTTTG-'3
Probe	5'-TGACGGGACTCCGCACAAGCG-'3	5'-ACACACCGCCCGTCGCTCC-'3
Reverse primer	5'-TCTTCGAATTAAACAACATGATCCA-'3	5'-AACACTTCACCGGACCATTCA-'3

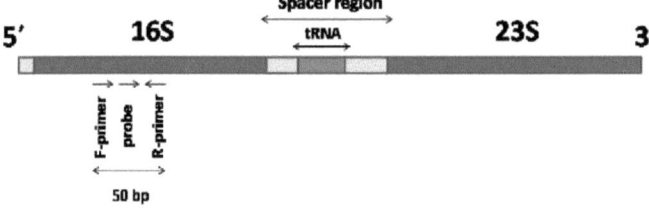

F-primer = forward primer for real time-PCR ; R-primer = reverse primer for real time-PCR

Figure.2.6.Diagrammatic representation of genomic location of qPCR primers and probe used for phytoplasma detection.

2.2.11. Transmission Electron Microscopy (TEM)

To demonstrate phytoplasmas directly the magnification and resolution of an electron microscope is required, due phytoplasmas are minute and lacking a defined shape. Preparation of thin sectioned resin-embedded samples and observations them by TEM enable both the revealing of the phytoplasma in the vascular tissues and studying the histological changes of the affected plants.

2.2.11.1. Preparation of thin sections

The resin used for embedding was similar to the one described by Spurr (1969).

Small pieces of leaf midribs of infected and healthy plants were fixed in 2% glutaraldehyde in 0.05 M phosphate buffer at 4°C overnight and then washed three times for 20 min in 0.05 M phosphate buffer. Tissues were then post-fixed in 2% osmium tetroxide (OsO4) in 0.05 M phosphate buffer at 4°C for overnight and then washed three times for 20 min in ddH_2O. Tissues were then dehydrated in a graded acetone series (25, 50, 70, 96, and 100%) and infiltrated in Spurr's low viscosity embedding medium (Spurr/ETOH 100%) as following:

1 part spurr to 3 parts ETOH for overnight
1 part spurr to 1 part ETOH for overnight
3 parts spurr to 1 part ETOH for overnight
100% spurr for overnight
Next polymerization as following:

40°C for 4 hours
50°C for 3 days

Ultrathin 60 nm sections were then cut with a diamond knife (type 35^0, Diatome, Biel, Switzerland) on a Leica UCT ultra cut microtome (Leica Microsystems, Wetzlar, Germany). Sections were post-stained for 10-15 min with 2 % Uranlylacetate in H_2O and for 8 min in lead citrate (Reynolds, 1963). Samples were examined in a ZEISS 902 (Zeiss, Oberkochem) electron microscope operated at 80 kV. Micrographs were taken using an Erlangshen ES500W CCD camera 1350 x 1050 pixel (Gatan, Peasanton CA) and Gatan Digital Micrograph software (Version 1.70.16).

3. Results

3.1. Establishment of the test for phytoplasma

"The polymerase chain reaction (PCR) incorporating mollicute-specific oligonucleotide primers derived from rRNA sequences made selective amplification of near full-length phytoplasma 16S rRNA genes from mixtures with host plant DNA possible. By its simplicity of application and superior sensitivity, PCR quickly became established as the method of choice for detection and diagnosis of phytoplasma diseases" (Mishra, 2004). Moreover, serology and DNA tests have been developed for diagnosis of phytoplasma diseases in sugarcane (Ratana, 2001; Srivastava et al., 2003). Of these techniques, PCR testing was found to be the most sensitive and reliable.

3.1.1. PCR for detection of phytoplasma

Because a single round of PCR was not able to detect low-titer phytoplasma infection, a second round of PCR was necessary. A nested PCR approach is often needed for detection of phytoplasmas (Schneider and Gibb, 1997) when they occur at low levels or are distributed unevenly in their plant hosts (Goodwin et al., 1994; Andersen et al., 1998). Poor amplification of target DNA by direct PCR is sometimes attributed to inhibitors present in host plant tissues (Cheung et al., 1993; Schneider and Gibb, 1997). Therefore the technique of nested PCR had been developed (Snounou et al., 1993; Kirkpatrick et al., 1994; Heinrich et al., 2001) in which a phytoplasma-specific stretch of 16S rRNA gene is amplified with primers in a first round, and then an internal part of this amplicon is amplified further in a second round with primers binding specifically to internal sequences of the first-round amplicon. Thus only the true positives which were generated by the first round are further multiplied, not however possible false positives.

In our lab, samples of DNA from phytoplasma-infected periwinkle (obtained from Dr. Seemüller, Dossenheim, Germany, and from Dr. Bertaccini, Bologna, Italy) were used as positive controls, whereas first round amplicon and second round amplicon of water control samples (instead of extracted DNA) were used as negative controls. The separation of the second round amplicons on agarose gels showed a clear band at the expected size of 1.2 kbp for the positive control using the primer pairs combination (P1/P7 and R16F2n/R16R2), and no bands for the water controls (Figure.3.1).

Results

In another experiment, DNA of the positive control was diluted by increasing quantities of DNA which had been extracted from sugarcane leaves of plants which were known to be phytoplasma-free. The purpose was to test, whether compounds from sugarcane leaves may possibly inhibit the amplification of phytoplasma 16S rRNA gene. The sugarcane extract by itself did not give an amplicon. The positive control sample always yielded a positive signal, even when diluted up to 40-fold by sugarcane DNA, the 50-fold dilution with sugarcane DNA did not give an amplicon anymore (Figure.3.1).

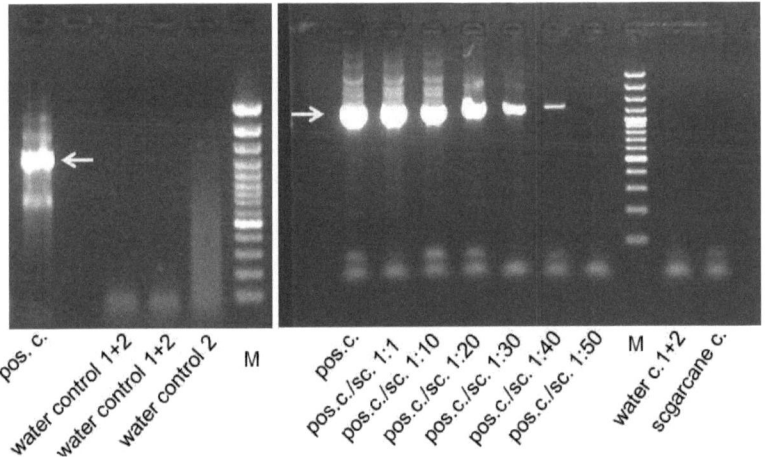

Figure.3.1. Nested PCR-products of positive and negative controls and of a positive control, which was mixed with increasing amounts of sugarcane DNA. The positive control (pos. c.) was a sample of American aster yellows phytoplasma grown in periwinkle (obtained from Dr. Bertaccini, Bologna). The water control 1+2 was with water instead of DNA in the first PCR round and further amplified in the second PCR round. Water control 2 contained water instead of first round amplicon. The sugarcane DNA was from Egyptian cultivar (Ph-8013)which had been shown to be phytoplasma-free (sugarcane c.) using the primer pairs combination (P1/P7 and R16F2n/R16R2). The marker (M) was DNA GeneRuler 100 bp plus (MBI Fermentas). The arrows point to the phytoplasma-specific band of 1.2 kbp.

Though PCR analysis is routine technique for phytoplasma detection, it's still meeting some difficulties, at least with some primers: several primer pairs and their combination are recommended (Heinrich et al., 2001). In our lab, PCR assay was carried out with different primer pairs combination. To amplify region that includes the 16S rRNA gene, the spacer region and the start of 23S rRNA gene of the phytoplasma genome. Therefore, each sugarcane

Results

sample was investigated for phytoplasma by using four nested-PCR assays which were numbering as following: (I), (II), (III) and (IV). The primer pairs and their sequences which used in each assay were mentioned at material and methods chapter.

3.1.2. Sources of sugarcane samples

Sugarcane samples, which were investigated in our lab, were obtained from different areas and different dates. Some of them were obtained as stem cuttings and grown in the greenhouse while others were harvested and conserved as air-dried leaves. Most of them are showing sugarcane yellow leaf syndrome symptoms whereas others were symptomless (Table.3.1).

Table.3.1. Original sources of sugarcane samples. Most of sugarcane samples were obtained from Hawaiian Islands while others from Thailand. In addition, some sugarcane samples were taken from Cuba and Middle East area including Egypt and Syria. Some of them were obtained as stem cuttings and grown in the greenhouse whereas others were collected and conserved as air-dried leaves.

Original source	Location	Date of getting sugarcane samples	enacragus fo epyT elpmas
iiawaH	Breeding station of HARC in Maunawili	2003	Stem cuttings
iiawaH	Plantations (Maui and Kauai)	2009	Sun-dried leaves
iiawaH	Former plantation fields (Maui, Kauai and Hawaii)	2009	Sun-dried leaves
iiawaH	Breeding station of HARC in Maunawili	2010	Sun-dried leaves
iiawaH	Close to former plantation fields in Hawaii	2011	Sun-dried leaves
iiawaH	Breeding station of HARC in Maunawili	2011	Sun-dried leaves
iiawaH	plantation (Maui)	2011	Sun-dried leaves
Thailand	Farmer fields (Bang Phra)	2010	Sun-dried leaves
Thailand	Breeding station (Khon Kean)	2010	Sun-dried leaves
Thailand	Farmer fields (Suphan Buri)	2011	Sun-dried leaves
abuC	Breeding station	2005	Stem cuttings
tpygE	Breeding station	2008	Stem cuttings
airyS	Farmer fields (Baniyas)	2008	Stem cuttings

3.2. Phytoplasma in sugarcane in Hawaii, Cuba, Egypt and Syria

Six sugarcane cultivars from Hawaii were obtained in 2003 as stem cuttings from the breeding station of HARC in Maunawili, Oahu , three SCYLV-susceptible cultivars (H87-4094, H73-6110, H65-7052) and three SCYLV-resistant cultivars (H78-7750, H78-4153, H87-4319).

These stem cuttings were grown in greenhouse of university of Bayreuth. In addition, cultivars from Cuba were obtained from Dr. Ortega, Habana, in 2005 also as stem cuttings and grown beside Hawaiian samples. Cultivars from Egypt were obtained as stem cuttings from the University of Gizah in 2008. The cultivar from Syria was obtained as stem cuttings in 2008 from a farmer's field near Baniyas. The question was as the following: Are these obtained sugarcane samples infected with phytoplasma?

3.2.1. Phytoplasma detection by nested-PCR assay (I) and identification by RFLP

DNA was extracted from source leaves and tested for phytoplasma by nested-PCR assay (I) with primer pairs (P1/P7 and R16F2n/R16R2) in 2008. All cultivars contained phytoplasma showing an amplicon at 1.2 kbp, although apparently at different titres, for example H73-6110, a strongly SCYLV-infected cultivar, had only a low SCYLP-titre (Figure.3.2). The Cuban cultivars (C10-5173, CP43-62, JA60-5) and one cultivar from Egypt (G84-47) was also infected by phytoplasma, although apparently at a low titre, not however the cultivar Gt-954 and Ph-8013 from Egypt and the plant from Syria (Figure.3.3). The results with cv. Gt-954, Ph-8013 and the Syrian cultivar thus were an important negative control, showing that there is no DNA sequence in the sugarcane genome which gives a false positive signal with this primer pair.

Results

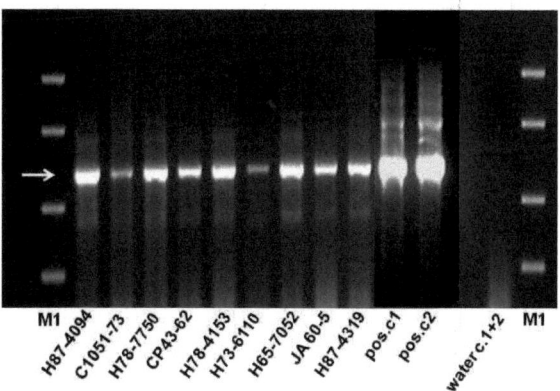

Figure.3.2. Phytoplasma in Hawaiian and Cuban sugarcane cultivars. DNA prepared from leaves of the indicated cultivars was tested with primers P1/P7 and R16F2n/R16R2. The positive control (pos. c1) was phytoplasma aster yellows from periwinkle obtained from Dr. Seemüller, Dossenheim, pos.c2 was phytoplasma aster yellows from periwinkle obtained from Dr. Bertaccini, Bologna. The water control 1+2 was with water instead of DNA in the first PCR round and further amplified in the second PCR round. The marker M1 is DNA ladder FastRuler Middle range (MBI Fermentas, fragment sizes: 4, 2, 1, 0.5 kbp). The arrows point to the phytoplasma-specific band of 1.2 kbp.

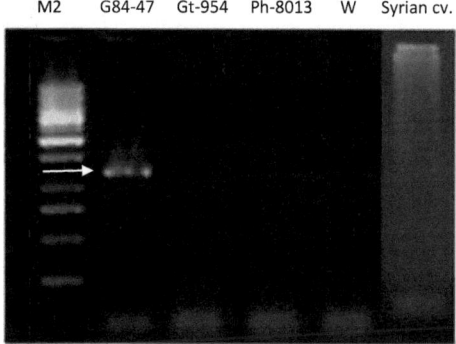

Figure.3.3. Phytoplasma in Egyptian and Syrian sugarcane cultivars. DNA prepared from leaves of the indicated cultivars was tested with primers P1/P7 and R16F2n/R16R2. Re-amplification of aliquot of first PCR water control with nested primer combination is in lane W. The marker M2 DNA GeneRuler 1kb (MBI Fermentas). The arrows point to the phytoplasma-specific band of 1.2 kbp.

Results

Restriction fragment analysis had been successfully applied to differentiate between the phytoplasma strains (Kirkpatrick et al., 1994; Lee et al., 1998; Valiunas et al., 2007).

The amplicons of the second round of PCR (I) were subjected to three restriction enzymes which were diagnostic for the phytoplasma strains. The obtained RFLP patterns were compared with those previously published by Lee et al., 1998.The restriction patterns identified the phytoplasma from Hawaiian cultivars and from one Cuban cultivar as belonging to the Aster yellows phytoplasma "*Ca.* Phytoplasma asteris", whereas the phytoplasma from the Cuban cultivar CP4362, which originally had been bred in Canal Point, Florida and from JA605, belonged to the Western X-disease phytoplasma "*Ca.* Phytoplasma pruni". However, a second profile was clearly visible in the gel in some Hawaiian sugarcane cultivars that indicate to the possible presence of phytoplasmas related to rice yellow dwarf group (16SrXI), "*Ca.* Phytoplasma oryzae" (Figure.3.4 and Table.3.2).

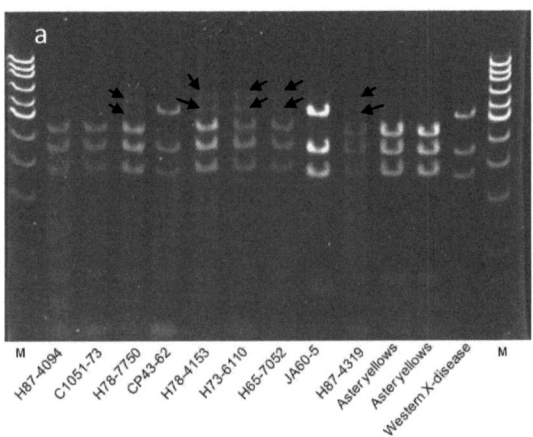

Results

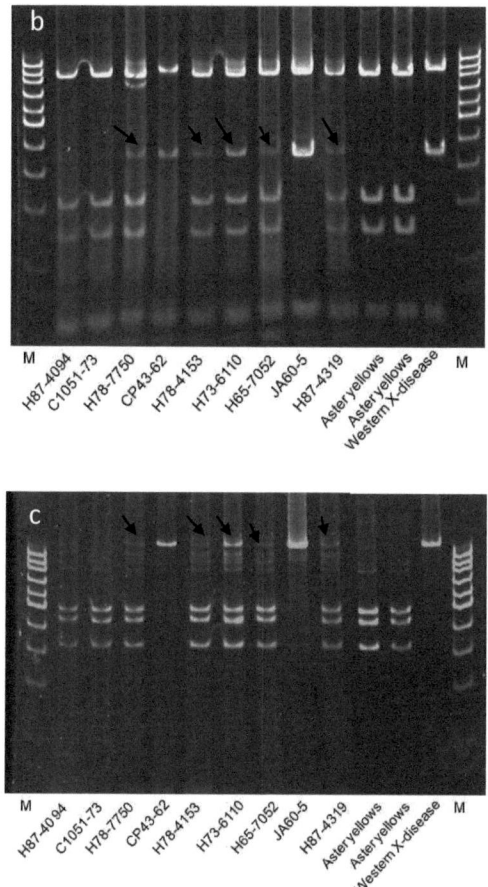

Figure.3.4.Restriction fragment analysis of PCR products from Hawaiian and Cuban sugarcane cultivars containing phytoplasma. The nested-PCR products were amplified with primers R16F2n/R16R2 following digestion with RsaI (a), HpaII (b) or KpnI (c) and separated on 5% polyacrylamide. The positive controls (Aster yellows and Western X-disease) were used as references. The black arrows indicate to second profile which reveals possible presence of rice yellow dwarf (16SrXI) phytoplasmas as mixed infection .The marker (M) is Mass Ruler DNA Ladder, low range (MBI Fermentas), fragment sizes 1031, 900,800,700, 600, 500, 400, 300, 200, 100, and 80 bp.

Table.3.2. Results of nested-PCR assay (I) and identification of phytoplasmas based on RFLP analyses.
Two phytoplasmas were identified in mixed infection in some Hawaiian sugarcane cultivars: one related to aster yellows group (16SrI) while the other tentatively related to rice yellow dwarf group (16SrXI). +, phytoplasma detected.

Sugarcane varieties	Original source	Detection of phytoplasma in 2008 based on PCR assay (I)	Phytoplasma group based on RFLP analyses of 16S rRNA gene
H87- 40 94	Hawaii	+	Aster yellows
C10- 51 73	Cuba	+	Aster yellows
H78- 77 50	Hawaii	+	Aster yellows and rice yellow dwarf
Cp- 43 62	Florida	+	X-disease
H78- 41 53	Hawaii	+	Aster yellows and rice yellow dwarf
H73- 61 10	Hawaii	+	Aster yellows and rice yellow dwarf
H65- 70 52	Hawaii	+	Aster yellows and rice yellow dwarf
JA- 60 5	Cuba	+	X-disease
H87- 43 19	Hawaii	+	Aster yellows and rice yellow dwarf
H87- 40 94 VF	Hawaii	+	Aster yellows

3.2.2. Phytoplasma detection by nested-PCR assay (II) and identification by RFLP

Oligonucleotide primers used for nested-PCR assay (II) were (SN910601/P6) for first-PCR and (R16F2n/R16R2) for nested-PCR. Hawaiian, Egyptian and Syrian sugarcane samples grown in greenhouse were tested by this PCR assay. According to this analyse four Hawaiian cultivars, two Egyptian cultivars and Syrian cultivar were negative for phytoplasma (Figure.3.5, Table.3.3 and 3.4).

Results

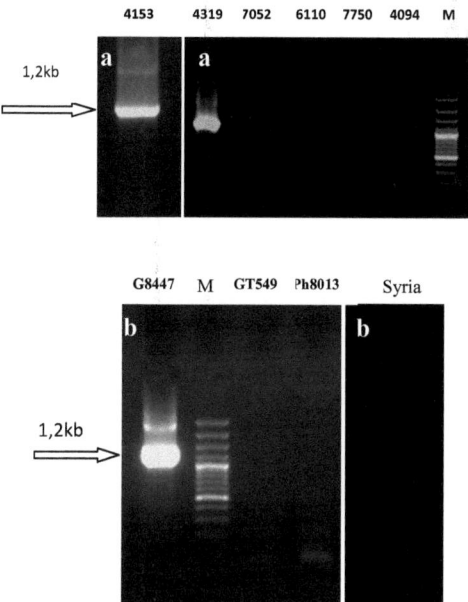

Figure.3.5.Nested-PCR assay (II) products (1.2kb) amplified with primers (SN910601/P6, R16F2n/R16R2).(a): Hawaiian sugarcane samples grown in greenhouse.**(b):** Egyptian and Syrian sugarcane samples. The marker M was GeneRuler 100 bp plus (MBI Fermentas). The arrows point to the phytoplasma-specific band of 1.2 kb. According to this analysis two Hawaiian sugarcane cultivars H78-4153 and H87-4319 and one Egyptian cultivarG8447 were positive for phytoplasmas.

Products of nested-PCR assay (II) were analyzed by RFLP analysis using single enzyme digestion with restriction endonucleases (HpaII and MseI). The obtained RFLP patterns were compared with those previously published by Lee et al., 1998. According to this digestion the Hawaiian cultivars H78-4153 and H87-4319 contain phytoplasmas fall in aster yellows group(Figure.3.6; aandTable.3.3) whereas Egyptian cultivarG8447 infected with phytoplasma belongs to rice yellow dwarf group (Figure.3.6; b and Table.3.4). However, further RFLP analysis is required to differentiate if this phytoplasma belongs to sugarcane white leaf strain (SCWL) or sugarcane grassy shoot one (SCGS).

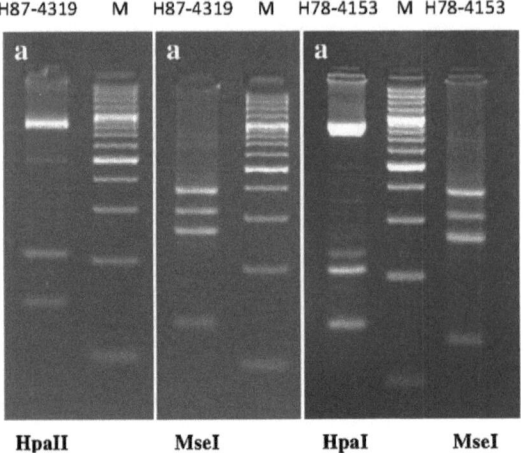

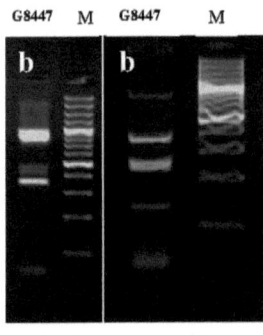

Figure.3.6.RFLP profiles of nested-PCR assay (II) products. These products amplified with primer pair (SN910601/P6, R16F2n/R16R2)of Hawaiian sugarcane samples (a) and Egyptian sugarcane samples (b) grown in greenhouse following single enzyme digestion with (HpaII and MseI) and separation on 2% agarose gel. The marker M was GeneRuler 100 bp plus (MBI Fermentas).According to this digestion the Hawaiian cultivars H78-4153 and H87-4319 contain phytoplasmas fall in aster yellows group while in Egyptian cultivar G8447 falls in rice yellow dwarf group.

3.2.3. Phytoplasma detection by nested-PCR assay (III)

Oligonucleotide primers used for nested-PCR assay III were (MLO-X/MLO-Y) for first-PCR and (P1/P2) for nested-PCR. According to this analyse all Hawaiian cultivars, one Egyptian (G8447) and Syrian cultivar were positive for the presence of phytoplasma but not the other two Egyptian cultivars (Figure.3.7, Table.3.3and 3.4).

Results

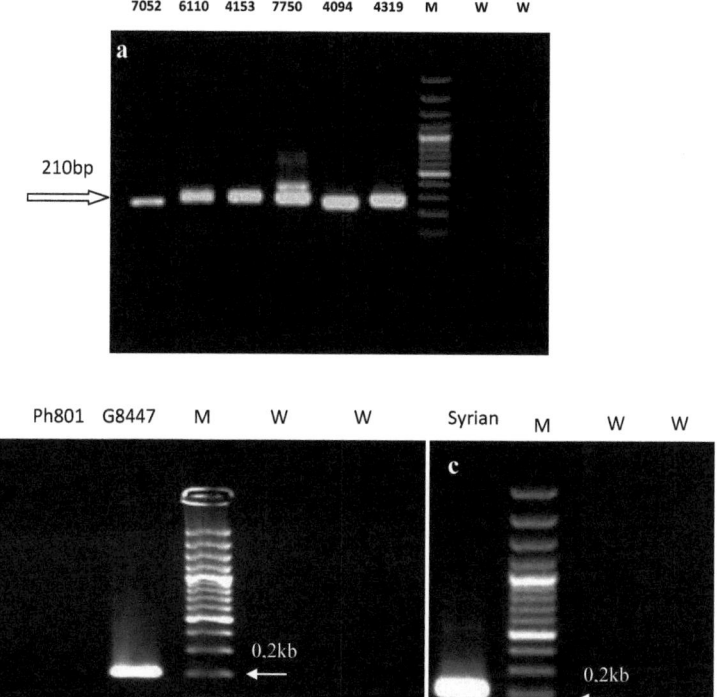

Figure.3.7. Nested-PCR assay (III) products (0.2kb)amplified with primer pair (MLO-X/MLO-Y, P1/P2). (a): Hawaiian sugarcane samples. (b): Egyptian sugarcane samples. (c): Syrian sugarcane sample grown in greenhouse. Re-amplification of aliquots of first PCR water controls with nested primer combination are in lanes W. The marker M was GeneRuler 100 bp plus (MBI Fermentas).The arrows point to the phytoplasma-specific band of 0. 2 kb.

3.2.4.Phytoplasma detection by nested-PCR assay (IV) and identification by RFLP

Oligonucleotide primers used for nested-PCR assay IV were (U-1/MLO-7) for first-PCR and (MLO-X/MLO-Y) for nested-PCR. According to these data the primers used in this assay could not detect phytoplasmal DNA present in all Hawaiian cultivars which were positive for phytoplasma as mentioned above (Table.3.3). Phytoplasmal DNA in Syrian and one Egyptian cultivar (G8447) was detected with this primer pair (Figure.3.8 and Table.3.4)

Results

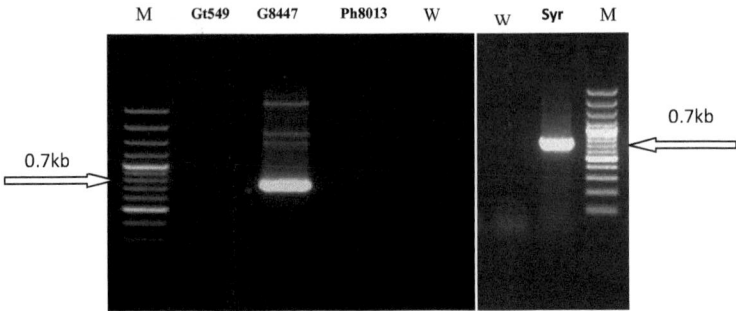

Figure.3.8. Nested-PCR assay (IV) products. Products (0.7kb) were amplified with primers (U-1/MLO-7, MLO-X/MLO-Y) obtained from Egyptian (Gt549, Ph8013 and G8447) and Syrian (Syr) sugarcane samples grown in greenhouse. Re-amplification of aliquot of first PCR water control with nested primer combination is in lane W. The marker M was GeneRuler 100 bp plus (MBI Fermentas).

Products of nested-PCR assay (IV) were analyzed by RFLP using single enzyme digestion with restriction endonucleas (HinfI). Rely on this RFLP analysis, phytoplasma strain of sugarcane white leaf (SCWL) can be differentiated from phytoplasma strain of sugarcane grassy shoot (SCGS) within the rice yellow dwarf group. The obtained RFLP patterns were compared with those previously published by Hanboonsong et al., 2002. According to this digestion the Egyptian cultivar (G8447) infected with phytoplasma strain of sugarcane grassy shoot (SCGS) but Syrian cultivar contains non-identified phytoplasma strain within the rice yellow dwarf group as indicated by DNA sequencing analysis (Figure.3.9 andTable.3.4).

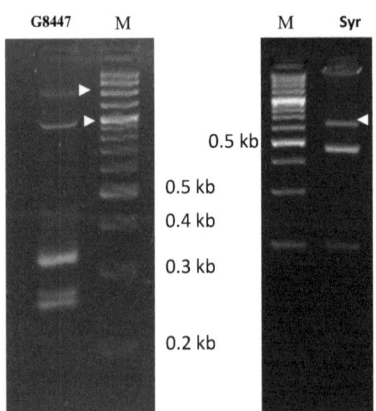

Figure.3.9. RFLP profiles of nested-PCR assay (IV) products. Products (0.7kb) amplified with primers (U-1/MLO-7, MLO-X/MLO-Y)of Egyptian (G8447) and Syrian (Syr) sugarcane samples following single enzyme digestion with (HinfI). The marker M was GeneRuler 100 bp plus (MBI Fermentas). Arrowheads indicate to non-specific bands. According to this digestion the Egyptian cultivar (G8447) infected with phytoplasma strain of sugarcane grassy shoot (SCGS)whereas Syrian one contentsnon-identified phytoplasma strain.

Results

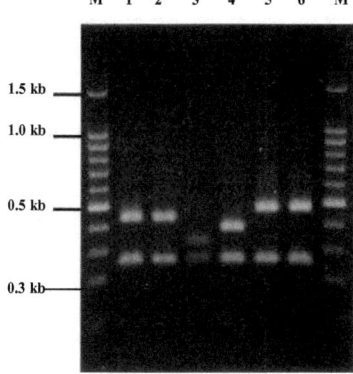

Figure.3.10. DNA amplified by nested-PCR with primers U-1/MLO-7 then primers MLO-X/MLO-Y and digested with Hinfl.1, phytoplasma from insect vector; 2, sugarcane white leaf; 3, sugarcane grassy shoot; 4, bermuda grass white leaf; 5, brachiaria grass white leaf; 6, crowfoot grass white leaf; M, 100 bp ladder. (Figure taken and adapted from Hanboonsong et al., 2002).

Table.3.3.Phytoplasma in Hawaiian sugarcane samples. Results of phytoplasma detection based on nested-PCR assays(II), (III) and (IV) of Hawaiian sugarcane samples grown in greenhouse and identification them based on RFLP analyses of nested-PCR assay(II) using single enzyme digestion with (HpaII and MseI). +,phytoplasma detected; -, phytoplasma not detected.

Sugarcane cultivar	phytoplasma detection based on PCR assay (II)	Phytoplasma group based on RFLP of PCR assay (II)	phytoplasma detection based on PCR assay (III)	phytoplasma detection based on PCR assay (IV)
H65-7052	-	-	+	-
H73-6110	-	-	+	-
H78-4153	+	Aster yellows	+	-
H78-7750	-	-	+	-
H87-4094	-	-	+	-
H87-4319	+	Aster yellows	+	-

Table.3.4. Phytoplasma in Egyptian and Syrian sugarcane samples. Results of phytoplasma detection based on nested-PCR assays (II), (III) and (IV) of Egyptian and Syrian sugarcane samples grown in greenhouse and identification them based on RFLP analysis of nested-PCR assay(II) using single enzyme digestion with (HpaII and MseI). And RFLP analysis of nested-PCR assay(IV) using single enzyme digestion with (HinfI) to differentiate rice yellow dwarf group to strains. +, phytoplasma detected; -, phytoplasma not detected.

Sugarcane cultivar	phytoplasma detection based on PCR assay (II)	Phytoplasma group based on RFLP of PCR assay (II) using (HpaII and MseI)	phytoplasma detection based on PCR assay (III)	phytoplasma detection based on PCR assay (IV)	Phytoplasma strain based on RFLP of PCR assay (IV)using (HinfI)
Gt549 (Egypt)	-	-	-	-	-
G8447(Egypt)	+	Rice yellow dwarf	+	+	Grassy shoot
Ph8013(Egypt)	-	-	-	-	-
Unknown cultivar (Syria)	-	-	+	+	unknown

3.3. Phytoplasma in sugarcane in Hawaiian plantations (2009)

As shown above, plants which were obtained from the Hawaiian breeding station were infected by phytoplasma of the Aster yellows and Rice yellow dwarf types. The plantations obtained the cultivars from the breeding station, followed by several cycles of field testing and multiplication. The question was whether the field plants also contained phytoplasma. Ratooning is not practiced in Hawaiian sugar industry; therefore the cultivars had been subjected to successive hot water treatments at each planting in the seed cane field and at each planting in the crop fields. However, different temperature regimes and durations are used for the hot water treatment, for example the HC&S plantation uses 50°C for 2 h for the seedcane field setts, and 52°C for 20 min for the crop field setts. According to our data, the latter may be too short to eliminate possible phytoplasma infection. Furthermore, it is unknown how much *de novo* infection occurred by phytoplasma vectors, e.g. leaf and plant hoppers, which are plenty in plantation fields. More than hundred plant hopper species have been described from the Hawaiian Islands (Asche, 1997), most of them endemic. Their potential to serve as phytoplasma vectors is unknown, however at least one leaf hopper species in Hawaii is known to transmit phytoplasma to water cress (Borth et al., 2006). Source leaves were collected from plantations from two islands (G&R in Kauai and HC&S in Maui). All samples were from cultivar H65-7052, a cultivar which is infected by SCYLV and expresses highly variable

Results

grades of YLS-symptoms. Symptomatic plants with yellow midribs may be found next to perfectly green plants without any obvious differences in soil and climate conditions. Therefore the possibility was envisaged that the symptomatic plants may contain phytoplasma in addition to SCYLV, either by insufficient thermotherapy or by *de novo* infection. Leaves from young plants and from adult plants of up to 20 months of age were tested. According to our investigation most of them contained phytoplasma (Table.3.5).

Plant material from the plantations had been sun-dried and later oven-dried to remove possible residual humidity. The intactness of the DNA was tested with the cytochrome oxidase (COX) gene as positive control. The primers for COX were COX-1 (5`- CCG GCG ATG ATA GGT GGA -`3) as forward primer and COX-2 (5`- GCC AGT ACC GGA AGT GA -`3) as reverse primer (the sequences were kindly provided by Dr. J. Hodgetts, Nottingham, UK). The PCR program was 95°C 4 min, (95°C 45 sec, 55°C 45 sec, 72°C 80 sec) x 30 cycles, 72°C 10 min. This generates an amplicon of approximately 400bp.

Figure.3.11shows that the control gene was clearly amplified from the DNA preparation of the dried leaves. There was plenty amplicon of the correct size, therefore the absence of the phytoplasma amplicon in some samples from the plantations cannot be accounted to destruction of DNA during the drying process.

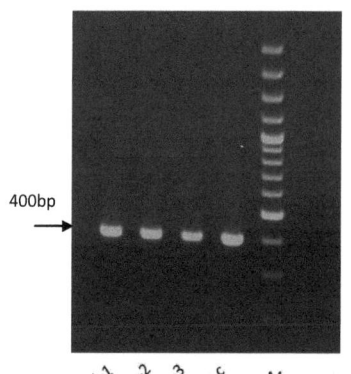

Figure.3.11. Amplification of the cytochrome oxidase (COX) sequence of DNA from dried leaves. To test whether the drying had destroyed too much DNA, the extracted DNA was also tested for cytochrome oxidase (COX) as control gene. DNA was extracted from sugarcane leaves and amplified by PCR to yield a 400 bp amplicon, which was separated on 1% agarose. The samples K1-1 to K1-3 from Kauai plantation, both air-dried and then oven-dried. The positive control (pos.c.) was from air-dried sugarcane leaves. The marker (M) was DNA GeneRuler 100 bp plus (MBI Fermentas).The arrow points to the COX-specific band of 400 bp.

Results

3.3.1. Phytoplasma detection and identification

Four nested-PCR assays (I), (II), (III) and (IV) were used to investigate the phytoplasma in sugarcane samples from Hawaiian plantation. However, no products were obtained except in PCR assay (III). The sequencing of some products showed the presence of rice yellow dwarf phytoplasma (Figure.3.12 and Table.3.5).

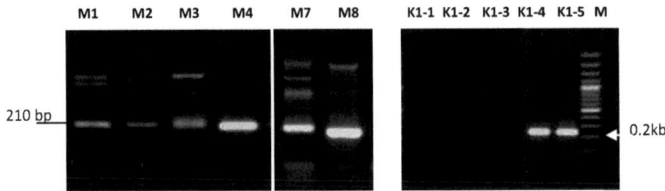

Figure.3.12. Phytoplasma in Kaui and Maui plantations sugarcane samples. Nested-PCR assay (III) products (0,2 kb) amplified with primers (MLO-X/MLO-Y, P1/P2). The samples M1 to M8 were from Maui and K1-1 to K1-5 from Kaui.

Table.3.5. Phytoplasma in Hawaiian plantations sugarcane samples (2009). Results of phytoplasma detection based on nested-PCR assays (I), (II), (III) and (IV) of sugarcane samples obtained from Hawaiian plantations as sun-dried leaves and identification them based on DNA sequencing analysis of nested-PCR assay (III) using the nested primer pair (P1/P2). According to the DNA sequencing analysis these sugarcane samples contain rice yellow dwarf phytoplasmas. +, phytoplasma detected; -, phytoplasma not detected. These samples were collected from plants of cultivar H65-7052. The samples (K1-1 to K1-5) from Kauai were collected from G&R plantation (4-5 km west of Olokele)while the samples (M1 to M8)from Maui were from HC&S plantation (fields 702, 500, 608 and 809, all south-east of Puunene). In each case 3 leaf samples from 3 different plants were tested.

Location, age of plant	Leaf condition	phytoplasma detection based on PCR assay (I), (II) and (IV)	phytoplasma detection based on PCR assay (III)	Phytoplasma group based on DNA sequencing of PCR assay (III) (P1/P2)
Kauai (H65-7052)				
3 months K1-1	Green leaves	-	-	
13 months K1-2	Green leaves	-	-	
" K1-3	YLS symptomatic	-	-	
23 months K1-4	Green leaves	-	+	Rice yellow dwarf
" K1-5	YLS symptomatic	-	+	Rice yellow dwarf
Maui (H65-7052)				
4 weeks M1	Green leaves	-	+	Rice yellow dwarf
3 months M2	Green leaves	-	+	Rice yellow dwarf
9 months M3	Green leaves	-	+	Rice yellow dwarf
" M4	YLS symptomatic	-	+	Rice yellow dwarf
20 months M7	Green leaves	-	+	Rice yellow dwarf
" M8	YLS symptomatic	-	+	Rice yellow dwarf

3.4. Phytoplasma in sugarcane in Hawaiian former plantation fields (2009)

Many Hawaiian sugarcane plantations went out of business in the past decades, but some sugarcane plants survived in the former sugarcane fields up to now due to the tropical climate and the perennial growth mode of sugarcane. These plants were standing on their places in the wild, in some cases since more than 30 years without replanting and thermotherapy (Komor et al. 2010) and they may have become infected by phytoplasma in case that the appropriate insect vectors were present. Samples of the uppermost fully unfolded leaves of plants found in former plantation fields were collected, sun-dried and tested for phytoplasma.

3.4.1. Phytoplasma detection and identification

The extracted DNA from these samples gave amplification only with primer pairs of PCR assay (III). The obtained amplicons were sequenced in order to identify the phytoplasma (Figure.3.13 and Table.3.6).

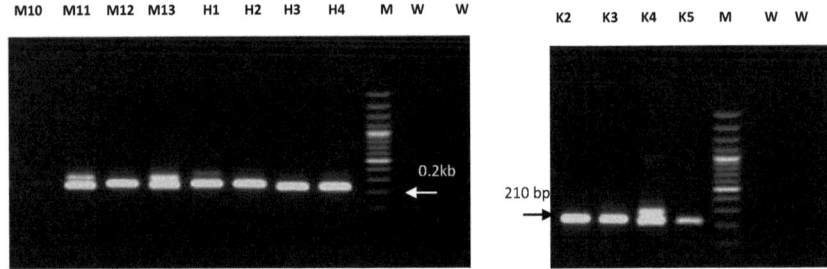

Figure.3.13. phytoplasma in former Hawaiian plantations sugarcane samples (2009). Nested-PCR assay (III) products (0,2kb) amplified with primers (MLO-X/MLO-Y, P1/P2). The samples M10 to M13 were from Maui plantation, H1 to H4 were from Hawaii plantation, K2 to K5 from Kauai, all as air-dried and then oven-dried. Re-amplification of aliquots of first PCR water controls with nested primer combinations are in lanes W. The marker (M) was DNA GeneRuler 100 bp plus (MBI Fermentas).

Results

Table.3.6. Phytoplasma in sugarcane samples from former Hawaiian plantations (2009). Results of phytoplasma detection based on nested-PCR assays and identification based on DNA sequencing of nested-PCR assay (III) using P1/P2. +, phytoplasma detected; -, phytoplasma not detected. The samples(H1 to H4) werefrom Hawaii while the samples(M10 to M13)were from Maui and the samples (K2 to K5) were from Kauai.

Island and collection site	phytoplasma detection based on PCR assay (I), (II) and (IV)	phytoplasma detection based on PCR assay (III)	Phytoplasma group based on DNA sequencing of PCR assay (III) (P1/P2)
Hawaii			
H1	-	+	Aster yellows
H2	-	+	Aster yellows
H3	-	+	Aster yellows
H4	-	+	Aster yellows
Maui (H65-7052)			
M10	-	+	
M11	-	+	Rice yellow dwarf
M12	-	+	Rice yellow dwarf
M13	-	+	Rice yellow dwarf
Kauai (H65-7052)			
K2	-	+	Aster yellows
K3	-	+	Aster yellows
K4	-	+	
K5	-	+	

3.5. Phytoplasma in sugarcane in Hawaiian breeding station (2010)

Six sugarcane cultivars from Hawaiian breeding station of HARC in Maunawili, Oahu were sent from Dr. Zhu in 2010 as sun-dried leaves in order to test for phytoplasmas.

3.5.1. Phytoplasma detection and identification

Only the primers of PCR assay (III) amplified DNA in these sugarcane cultivars. However, the obtained bands were very weak, therefore; only two of them were sequenced (Figure.3.14andTable.3.7).

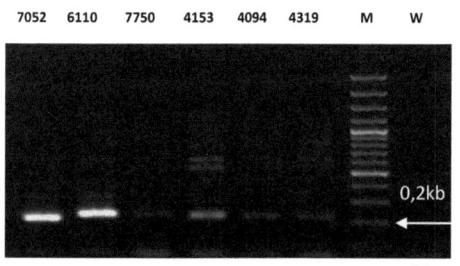

Figure.3.14. phytoplasma in Hawaiian breeding stationsugarcane samples (2010). Nested-PCR assay (III) products (0.2kb) amplified with primers (MLO-X/MLO-Y, P1/P2) obtained from Hawaiian sugarcane breeding station (2010) as air-dried samples. Re-amplification of aliquot of first PCR water control with nested primer combination is in lane W. The marker (M) was DNA GeneRuler 100 bp plus (MBI Fermentas).

Results

Table.3.7. Phytoplasma in sugarcane plants from Hawaiian (Maunawili, HARC)breeding station(2010).
Results of phytoplasma detection based on nested-PCR assays and identification by DNA sequencing of nested-PCR assay (III) using P1/P2. +, phytoplasma detected; -, phytoplasma not detected.

Sample name	phytoplasma detection based on n-PCR assay (I)& (II)& (IV)	phytoplasma detection based on n-PCR assay (III)	Phytoplasma group based on DNA sequencing (P1/P2)
H65-7052	-	+	Rice yellow dwarf
H73-6110	-	+	Aster yellows
H78-7750	-	+	
H78-4153	-	+	
H87-4094	-	+	
H87-4319	-	+	

3.6. Phytoplasma in sugarcane in Hawaiian breeding station and plantations (2011)

In February 2011, sugarcane leaf samples were harvested from different areas in Hawaiian Islands including plantation HC&S, Maui and Maunawili breeding station. Then, these sugarcane samples were sun-dried in order to test the presence of phytoplasma in our lab. Most of these samples were taken from sugarcane plants are showing sugarcane yellow leaf syndrome symptoms (Figure.3.15).

3.6.1. Phytoplasma in sugarcane in Hawaiian plantation

These plantations obtained the cultivars from the breeding station of HARC in Maunawili, Oahu, followed by several cycles of field testing and multiplication. The question was if phytoplasmas can be responsible for YLS in this plantation field and if there is significant correlation between the presence of phytoplasma and showing sugarcane yellow leaf syndrome since some samples were strongly or slightly symptomatic while others asymptomatic. Samples from uppermost fully unfolded source leaves from plants of different cultivars were collected in the plantation fields and sun-dried. The samples from Maui were from HC&S plantation (fields 702, 500, 608 and 809, all south-east of Puunene). In each case 3 leaf samples from 3 different plants were tested.

Results

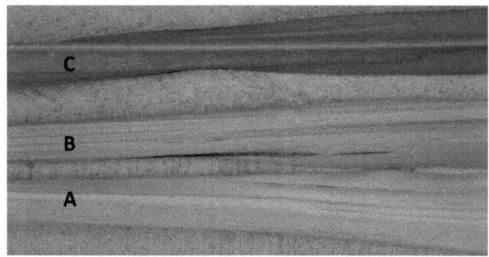

Figure.3.15. Sugarcane leaves showing symptoms of infection with yellow leaf syndrome (A) and (B), compared with an uninfected green leaf (C). Sugarcane yellow leaf syndrome symptoms are caused by several agents including phytoplasma.

3.6.1.1. Phytoplasma detection and identification

DNA was extracted from source leaves and each sugarcane sample was investigated for the presence of phytoplasma by using four nested-PCR assays (I), (II), (III) and (IV) with different primer pair combinations as was clarified before. Only nested-PCR assay (III)gave positive reactions. Our results showed that symptomatic and non- symptomatic plants contain phytoplasma ((Figure.3.16 and Table.3.8).

Results

Table.3.8. Phytoplasma in sugarcane plants from Hawaiian HC&Splantation of Maui island, close to Puunene (2011). Results of phytoplasma detection based on nested-PCR assays and identification based on DNA sequencing of products of nested-PCR assay (III) using P1/P2. According to the DNA sequencing analysis the sugarcane samples contain rice yellow dwarf phytoplasma. +, phytoplasma detected; -, phytoplasma not detected.

Order of Sugarcane varieties	Sugarcane cultivar and Leaf condition	phytoplasma detection based on PCR assay (I), (II), (IV)	phytoplasma detection based on PCR assay (III)	Phytoplasma group based on DNA sequencing (P1/P2)
1	H65-7052, 6 months Non-symptomatic	-	+	
2	H65-7052, 6 months symptomatic	-	-	-
3	H73-3567, 4 months Non-symptomatic	-	-	-
4	H87-4319, 9 months Non-symptomatic	-	+	Rice yellow dwarf
5	H87-4319, 9 months slightly-symptomatic	-	+	
6	H86-3792, 6 months Non-symptomatic	-	+	Rice yellow dwarf
7	H87-5794, 9 months Non-symptomatic	-	+	Rice yellow dwarf

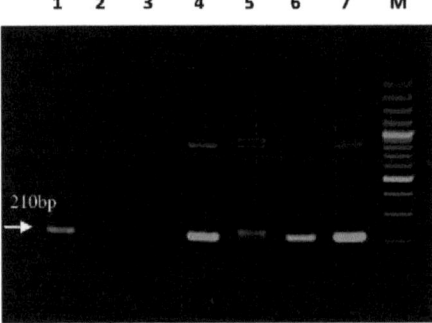

Figure.3.16. Phytoplasma in Hawaiian HC&Splantation of Mauiisland, close to Puunene(2011). Nested-PCR assay (III) products (0.2kb) amplified with primers (MLO-X/MLO-Y, P1/P2) of sugarcane samples obtained from Hawaiian plantation HC&S, Maui, close to Puunene; (2011) as sun-dried leaves. The marker (M) was DNA GeneRuler 100 bp plus (MBI Fermentas).

Results

3.6.2. Phytoplasma in sugarcane in Hawaiian breeding station (2011)

Samples from uppermost fully unfolded source leaves from 10 cultivars of sugarcane plants were collected from HARC breeding station in Maunawili(fields A, B and P which are widely distant from each other)and sun-dried for phytoplasma investigation. Names of these sugarcane cultivars are indicated in (Table.3.9).These cultivars were previously tested for sugarcane yellow leaf virus (SCYLV) by Lehrer et al., 2001.The results of this test are indicated in (Table.3.11).

3.6.2.1. Phytoplasma detection by nested-PCR assay (II) and identification

DNA was extracted and tested for phytoplasma by nested-PCR assay (II).Most of these sugarcane cultivars produced an amplicon at 1.2 kbp, although apparently at different titres (Figure.3.17). Products of nested-PCR assay (II) were digested with restriction endonucleases (HpaII and MseI). The obtained RFLP patterns were compared with those previously published by Lee et al., 1998. According to this digestion these Hawaiian sugarcane samples contain phytoplasmas fall in rice yellow dwarf group (Figure.3.18 andTable.3.9). However, further RFLP analysis is required to differentiate if this phytoplasma belongs to sugarcane white leaf strain (SCWL) or sugarcane grassy shoot one (SCGS) as mentioned before.

Table.3.9. Phytoplasma in Hawaiian (Maunawili, HARC) breeding station sugarcane samples (2011). Results of phytoplasma detection based on nested-PCR assay (II) and identification based on RFLP analysis using single enzyme digestion with HpaII and MseI. Sample numbers indicate to the order of samples in next figures (3.17) and (3.18). +, phytoplasma detected; -, phytoplasma not detected.

Samples number	Samples name	phytoplasma detection based on PCR assay (II)	Phytoplasma group based on RFLP With HpaII and MseI
1	H87-4094 field A11	+	
2	H78-3567 "	+	Rice yellow dwarf
3	H87-4319 "	+	
4	H65-7052 "	+	
5	H50-7209 "	+	Rice yellow dwarf
6	H78-4153 "	+	Rice yellow dwarf
7	H77-4643 "	+	Rice yellow dwarf
8	H73-6110, field A22	+	
9	H32-8560, field B31		
10	H78-7750, field B62	+	
11	H87-4319 "	+	Rice yellow dwarf
12	H78-3606 "	-	-
13	H77-4643, field P11a	+	
14	H78-3606, field P11	-	-
15	H50-7209, field P12	-	-
16	H87-4319 "	+	
17	H65-7052 "	-	-
18	H78-7750 "	+	
19	H73-6110, field P13	+	

Results

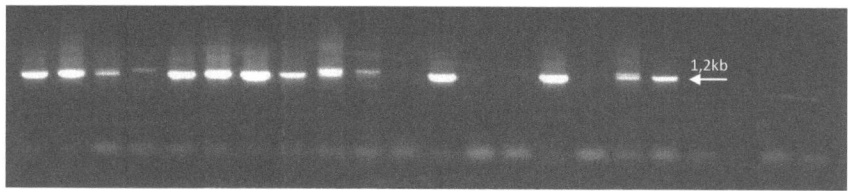

Figure.3.17. Phytoplasma detection in sugarcane samples from Hawaiian (Maunawili, HARC) breeding station (2011).Nested-PCR assay (II) products (1.2kb) amplified with primers (SN910601/P6, R16F2n/R16R2). Samples numbers and names are indicated in table (3.9) above. Re-amplification of aliquots of first PCR water controls with nested primer combinations are in lanes W. The marker (M) was DNA GeneRuler 100 bp plus (MBI Fermentas).

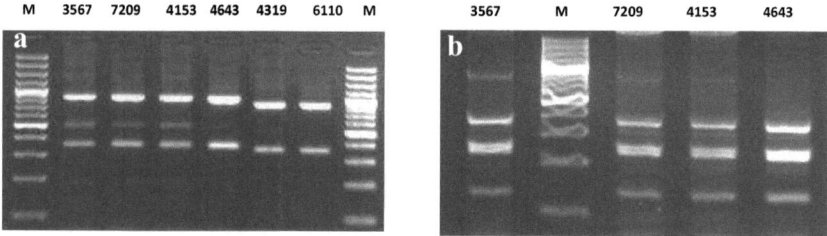

Figure.3.18. Phytoplasma identification in sugarcane samples from Hawaiian (Maunawili, HARC) breeding station(2011).RFLP profiles of nested-PCR assay (II) products (1.2kb) amplified with primers (SN910601/P6, R16F2n/R16R2) following single enzyme digestion with HpaII (a) and MseI (b). The marker M was GeneRuler 100 bp plus (MBI Fermentas).

3.6.2.2. Phytoplasma detection by nested-PCR assay (III) and (IV) and identification

According to PCR assay (III) results most of these sugarcane cultivars were positive for phytoplasma (Figure.3.19). This is also true for PCR assay (IV) (Figure.3.20). Products of PCR assay (IV) were digested with restriction endonuclease (HinfI).According to this digestion these Hawaiian sugarcane samples contain phytoplasma strain of sugarcane white leaf (SCWL) (Figure.3.21andTable.3.10).

Table.3.10. Phytoplasma in sugarcane samples from Hawaiian (Maunawili, HARC) breeding station (2011). Results of phytoplasma detection based on nested-PCR assays (III) and (IV) and identification based on RFLP analysis of (IV) products using single enzyme digestion with (HinfI). Sample numbers indicate to the order of samples in next figures (3.19 and 3.20). +, phytoplasma detected; -, phytoplasma not detected.

Sample number	Sample name	Phytoplasma detection based on PCR assay (III)	Phytoplasma detection based on PCR assay (IV)	Phytoplasma strain based on RFLP with (HinfI)
1	H87-4094 field A11	+	+	Sugarcane white leaf
2	H78-3567	+	+	"
3	H87-4319	+	+	"
4	H65-7052	+	+	"
5	H50-7209	+	+	"
6	H78-4153	+	+	"
7	H77-4643	-	+	"
8	H73-6110 field A22	+	+	"
9	H32-8560 field B31	-	-	-
10	H78-7750 field B62	+	+	"
11	H87-4319	-	-	-
12	H78-3606	+	+	"
13	H77-4643 field P11a	+	+	"
14	H78-3606 field P11	+	+	"
15	H50-7209 field P12	-	+	-
16	H87-4319	+	-	"
17	H65-7052	-	+	-
18	H78-7750	+	+	"
19	H73-6110	+	+	"

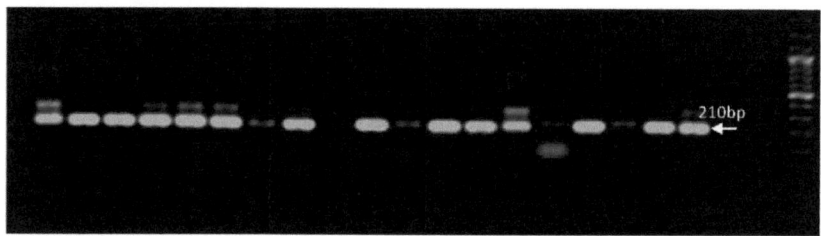

Figure.3.19. Phytoplasma detection in sugarcane samples from Hawaiian (Maunawili, HARC) breeding station sugarcane samples (2011). Nested-PCR assay (III) products (0.2kb) amplified with primers (MLO-X/MLO-Y, P1/P2). Samples numbers and names are indicated in (Table.3.10). Re-amplification of aliquots of first PCR water controls with nested primer combination is in lanes W. The marker (M) was DNA GeneRuler 100 bp plus (MBI Fermentas).

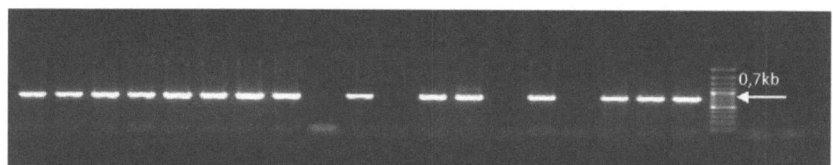

Figure.3.20. Phytoplasma detection in sugarcane samples from Hawaiian (Maunawili, HARC) breeding station (2011). Nested-PCR assay (IV) products (0.7kb) amplified with primers (U-1/MLO-7, MLO-X/MLO-Y). Samples numbers and names are indicated in (Table.3.10). Re-amplification of aliquots of first PCR water controls with nested primer combination is in lanes W. The marker (M) was DNA GeneRuler 100 bp plus (MBI Fermentas).

Results

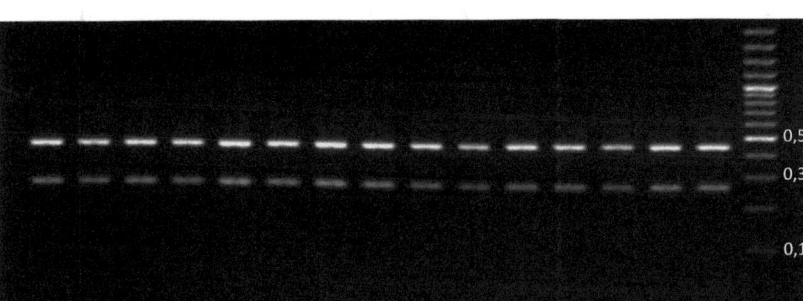

Figure.3.21. Phytoplasma identification in sugarcane samples from Hawaiian (Maunawili,HARC) breeding station(2011). RFLP profiles of nested-PCR assay (IV) products (0.7kb) amplified with primers (U-1/MLO-7, MLO-X/MLO-Y) following single enzyme digestion with (HinfI). The marker M was GeneRuler 100 bp plus (MBI Fermentas). RFLP patterns were compared with those previously published by Hanboonsong et al., 2002.

Table.3.11. Presence of SCYLV in sugarcane cultivars in the HARC breeding station in 2001 and presence of phytoplasma in these cultivars which collected in (2011). SCYLV was assayed by tissue blot immunoassay but the presence of phytoplasma was tested by nested-PCR.

sravitluc enacragus	SCYLV detection based on Tissue blot immunoassay	phytoplasma detection based on nested-PCR assay
H87-4094	+	+
H78-3567	+	+
H87-4319	+	+
H65-7052	+	+
H50-7209	+	+
H78-4153	-	+
H77-4643	+	+
H73-6110	+	+
H78-7750	-	+
H78-3606	+	+

Results

3.6.3. Phytoplasma in sugarcane in different areas close to former plantations

Original sources of these sugarcane samples are indicated in (Table.3.12). All these collected sugarcane samples should show whether sugarcane plants in the neighborhood of water cress farming have phytoplasma. Except the sample from garden of Dr. Lehrer and samples from Akaka Falls (300 m and 200 m elevation) should show whether phytoplasma may be responsible for YLS, since the virus titre was absent or low.

3.6.3.1. Phytoplasma detection and identification

The extracted DNA from these samples gave amplification with primer pairs of PCR assay (III) and (IV). The obtained amplicons were sequenced and digested with restriction endonuclease (HinfI). According to this digestion one sugarcane cultivar contains sugarcane white leaf phytoplasma strain whereas other cultivar infected with un-identified strain within rice yellow dwarf group (Figure.3.23; b and Table.3.12).

Table.3.12. Phytoplasma in Hawaiian sugarcane samples from different sources close to former plantations (2011). Results of phytoplasma detection based on nested-PCR assays and identification them based DNA sequencing and RFLP analysis of (IV) products using single enzyme digestion with (HinfI). Sample numbers indicate to the order of samples in next figures (3.22 and 3.23). +, phytoplasma detected; -, phytoplasma not detected.

Sample number	Sample name	phytoplasma detection based on PCR assay (II)	phytoplasma detection based on PCR assay (III)	phytoplasma detection based on PCR assay (IV)	Phytoplasma strain based on RFLP with (HinfI)
1	Virus-free from. Lehrer	-	+ Rice yellow dwarf	+	Unknown
2	Honomu, Stable Camp Rd	-	+ Rice yellow dwarf	+	Sugarcane white leaf
3	Honomu, Akaka Falls Rd	-	+ Rice yellow dwarf	-	-
4	Honomu, Akaka Falls Rd	-	+ Rice yellow dwarf	+	
5	Akaka Falls, 300m elevation	-	+ Rice yellow dwarf	-	-
6	" , 200m elevation	-	-	-	-
7	Kukui Camp	-	+ Rice yellow dwarf	-	
8	Keanae, Hana Rd., close to Taro	-	+ Rice yellow dwarf	-	

Results

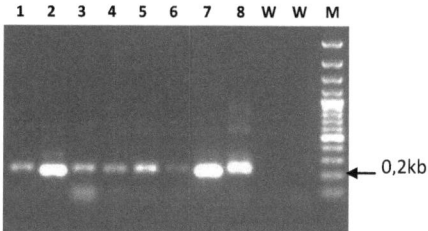

Figure.3.22. Phytoplasma detection in Hawaiian sugarcane samples from different sources close to former plantations (2011). Nested-PCR assay (III) products (0.2kb) of sugarcane samples obtained from different sources close to former plantations as sun-dried leaves. Samples numbers and names are indicated in (Table.3.12). Re-amplification of aliquots of first PCR water controls with nested primer combinations are in lanes W. The marker (M) was DNA GeneRuler 100 bp plus (MBI Fermentas).

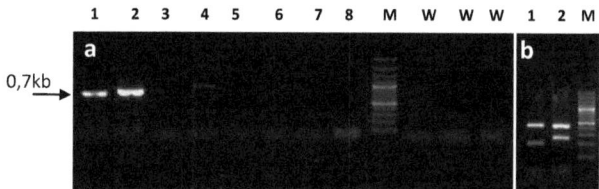

Figure.3.23. Phytoplasma in Hawaiian sugarcane samples from different sources close to former plantations (2011). (a): Nested-PCR assay (IV) products (0.7kb) sugarcane samples obtained from different sources close to former plantations in (2011) as sun-dried leaves. Samples numbers and names are indicated in table 11. Re-amplification of aliquots of first PCR water controls with nested primer combination is in lanes W. The marker (M) was DNA GeneRuler 100 bp plus (MBI Fermentas).(b): RFLP profiles of nested-PCR assay (IV) products following single enzyme digestion with (HinfI). RFLP patterns were compared with those previously published by Hanboonsong et al., 2002.

3.6.4. Phytoplasma in grass weeds in Sumida watercress farm

Further surveys should be done in order to check whether sugarcane plants which are growing close to water cress fields contain the water cress-typical phytoplasma. However, Sumida watercress field in Pearlridge had phytoplasma-infected water cress 10 years ago and these plants had been eliminated. Furthermore, currently there are no sugarcane plants nearby this field. However, some of perennial grasses (some of them are wild relatives of sugarcane such as (*Miscanthus Sp.*) are still present nearby this field which may be reservoir of phytoplasma. Therefore, perennial grasses were collected from Sumida water cress field and were brought to our lab in order to test whether these perennial grasses are infected with phytoplasma. According to our investigation, however, these grasses are not infected with phytoplasma (Table.3.13).

Table.3.13. Outline the detection of phytoplasmas; in samples were obtained from grass weeds from Sumida watercress farm in pearlidge; based on nested-PCR assays.

Grasses sample	PCR assay(II)	PCR assay (III)	PCR assay (IV)
Paspalum coniugatum	-	-	-
Cyperus rotundus	-	-	-
Panicum maximum	-	-	-
Miscanthus Sp.	-	-	-

3.7. Phytoplasma in sugarcane in Thailand (2010 – 2011)

Sugarcane white leaf (SCWL) is one of the most destructive sugarcane diseases in Thailand. It was first observed in 1954 in the Lumpang province in the northern part of Thailand (Mangelsdorf, 1962). Only four years later, SCWL was discovered in Taiwan (Ling, 1962).

Sugarcane white leaf (SCWL) and sugarcane grassy shoot (SCGS) have been associated with distinct phytoplasma strains within the rice yellow dwarf taxonomic group (16SrXI), "*Candidatus Phytoplasma oryzae*" that causes rice yellow dwarf disease (Marcone et al., 2004, Ariyarathna et al., 2007). SCWL phytoplasma appears to be more closely related to SCGS phytoplasma than to phytoplasmas causing white leaf symptoms in some grasses. Furthermore, SCWL and SCGS phytoplasmas could be differentiated by RFLP analysis of rRNA gene using suitable restriction endonucleases (Marcone, 2002).

3.7.1. Phytoplasma in sugarcane in Bang Phra and Khon Kean provinces (2010)

In October 2010, sugarcane leaf samples were harvested from Bang Phra and Khon Kean (KK) provinces in Thailand including farmer fields (F) and breeding stations (S). Then, these sugarcane samples were sun-dried and were brought to our lab in order to investigate for phytoplasma. These samples were taken from sugarcane plants are showing different diseases symptoms including sugarcane white leaf syndrome symptoms (Figure.3.24). Names of these sugarcane samples and their diseases symptoms are indicated in (Table.3.14).

Results

Figure.3.24. Sugarcane leaves samples from Thailand. These leaves were collected from farmer fields (F) and breeding station (S) in province of Bang Phra in Thailand and used for phytoplasma investigation. Bleaching leaves at the right side show sugarcane white leaf symptoms whereas the red one in the middle shows the rust disease symptoms beside it at the left hand leaf shows yellow spots. Two stunted leaves are at the left side of bleaching leaves while the last two leaves at the left side are infected with curly spindle disease.

3.7.1.1. Phytoplasma detection and identification

The extracted DNA from these samples gave amplification with primer pairs of PCR assay (III) and (IV) (Figure.3.25 and.3.26). Products of assay (IV) were digested with restriction endonuclease (HinfI). According to this digestion phytoplasma infected-sugarcane samples contain sugarcane white leaf phytoplasma strain (Figure.3.27 and Table.3.14).

Results

Table.3.14. Phytoplasmas in sugarcane samples from provinces of Bang Phra and Khon Kean in Thailand in 2010; based on nested-PCR assays. Samples (F1 to F8) were from farmer fields in province of Bang Phra while the samples (S1 to S13) were from breeding station also in province of Bang Phra. Samples (KK1 to KK38) were from province of Khon Kean.

Sugarcane sample	Desiease Symptoms	PCR assay (III)	PCR assay (IV)	Phytoplasma strains based on RFLP with (HinfI)
F1	White fly	+	+	Sugarcane white leaf
F2	Leaf spot	+	+	"
F3	rust	+	+	"
F4	Yellow spot	+	+	"
F5	Mosaic virus	+	+	"
F6	Curly spindle	+	+	"
F7	Stunted leaf	+	+	"
F8	White leaf	+	+	"
S1	Spotted mosaic	-	-	
S2	Spotted mosaic	+	+	Unknown
S3		-	-	
S4	rust	-	+	Unknown
S5	rust	-	+	Unknown
S6		-	-	
S7		-	-	
S8		-	-	
S9		+	-	
S10	Streak mosaic	+	+	Sugarcane white leaf
S11	Streak mosaic	+	+	Unknown
S12	Streak mosaic	-	-	
S13	Streak mosaic	-	-	
KK1		+	+	Sugarcane white leaf
KK2	Mosaic	+	+	"
KK3	Grassy shoot	+	+	"
KK4	Yellow midrib	+	+	"
KK11	Yellow midrib	+	-	
KK12	Leaf scalel	+	+	"
KK13	Erianthus cross yellow midrib	+	-	
KK14	mosaic	+	+	"
KK17	mosaic	+	+	"
KK18	Erianthus	+	-	
KK20	mosaic	+	+	"
KK21	Spot	-	-	
KK32	white leaf	+	+	"
KK33	white leaf	+	+	"
KK34	white leaf	+	+	"
KK35	Stripe	+	+	"
KK36	Stripe	+	+	"
KK37	Yellow midrib	+	+	"
KK38	Yellow midrib	+	+	"

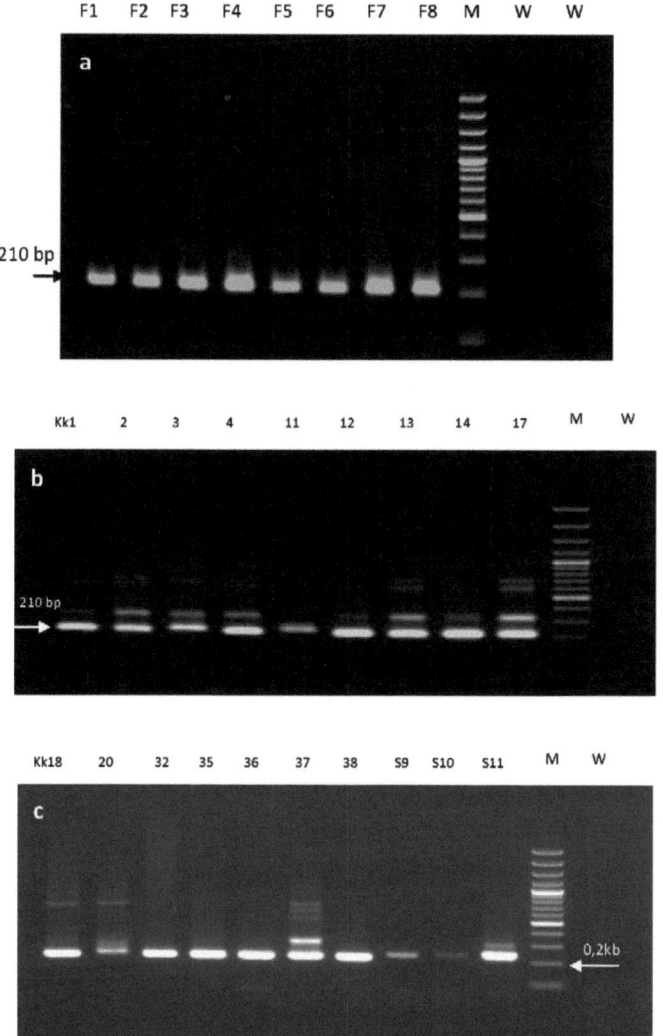

Figure.3.25; a, b and c. Phytoplasma detection in Thai sugarcane samples from Bang Phra and Khon Kean. Nested-PCR assay (III) products (0.2kb) of sugarcane samples obtained from farmer fields (F1 to F8) and breeding station (S9 to S11) in Bang Phra and Khon Kean (KK1 to KK38) as sun-dried leaves. Samples numbers and names are indicated in (Table 3.14). Re-amplification of aliquots of first PCR water controls with nested primer combinations are in lanes W. The marker (M) was DNA GeneRuler 100 bp plus (MBI Fermentas).

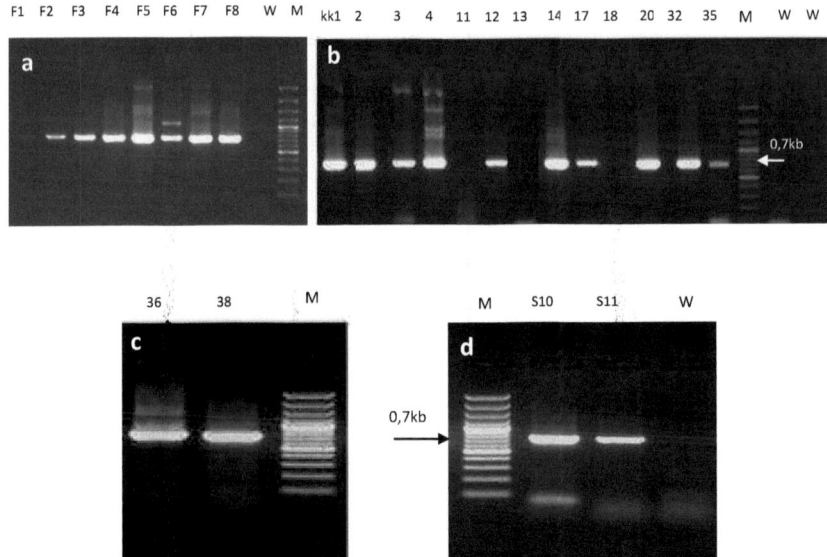

Figure.3.26; a, b, c and d. Phytoplasma detection in Thai sugarcane samples from Bang Phra and Khon Kean. Nested-PCR assay (IV) products (0.7kb) of sugarcane samples obtained from farmer fields (F1 to F8) and breeding station (S10 to S11) in Bang Phra and Khon Kean (KK1 to KK38) as sun-dried leaves. Samples numbers and names are indicated in (Table 3.14). Re-amplification of aliquots of first PCR water controls with nested primer combination is in lanes W. The marker (M) was DNA GeneRuler 100 bp plus (MBI Fermentas).

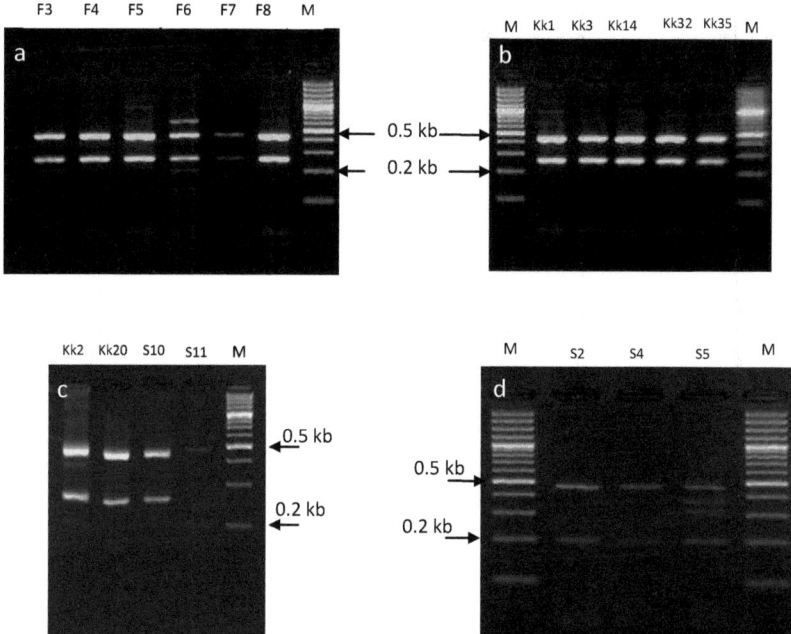

Figure.3.27; a, b, c and d. **Phytoplasma identification in Thai sugarcane samples from Bang Phra and Khon Kean.** RFLP profiles of nested-PCR assay (IV) products (0.7kb) of sugarcane samples obtained from farmer fields (F3 to F8) and breeding station (S2 to S11) in Bang Phra and Khon Kean (KK1 to KK35) as sun-dried leaves following single enzyme digestion with (HinfI). The marker M was GeneRuler 100 bp plus (MBI Fermentas).

3.7.2. Phytoplasma in sugarcane in Suphan Buri province (2011)

Sugarcane samples were sent from farmer fields in province of Suphan Buri as sun-dried leaves. Four samples were taken from sugarcane plants showing yellow leaf syndrome symptoms and one sample was taken from plant shows sugarcane grassy shoot symptoms.

3.7.2.1. Phytoplasma detection and identification

The extracted DNA from these samples gave amplification only with primer pairs of PCR assay (III). Some products of this assay were sequenced for the identification (Figure.3.28 and Table.3.15).

Results

Table.3.15. Outlines of phytoplasmas in Thai sugarcane samples from Suphan Buri.

Sugarcane sample	phytoplasma detection based on PCR assay (III)	phytoplasma detection based on PCR assay (IV)	Phytoplasma group based on DNA sequencing (P1/P2)
1	+	-	Rice yellow dwarf
2	+	-	,,
3	+	-	"
4	+	-	"
Grassy shoot	+	-	"

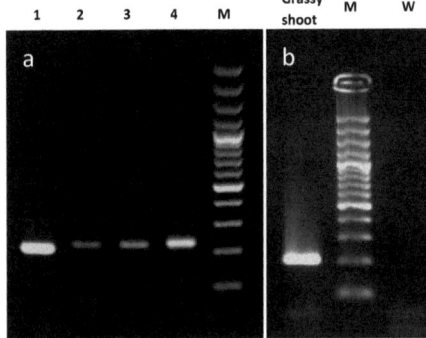

Figure.3.28; a and b.Phytoplasma in Thia sugarcane samples from Suphan Buri. Nested-PCR assay (III) products (0.2kb) of sugarcane samples obtained from farmer fields in Suphan Buri as sun-dried leaves. Samples numbers and names are indicated in (Table.3.15). Re-amplification of aliquot of first PCR water control with nested primer combination is in lane W. The marker (M) was DNA GeneRuler 100 bp plus (MBI Fermentas).

3.8. Establishment of TaqMan qPCR assay as another test for phytoplasma

Most universal as well as specific phytoplasma diagnostic protocols rely on nested PCR, which, although extremely sensitive, is also time-consuming and possess risks in terms of carry-over contamination between the two rounds of amplification (Weintraub and Jones, 2010). Recently, direct qPCR has replaced the traditional PCR in efforts to increase the speed and sensitivity of detection and to improve techniques for mass screening (Weintraub and Jones, 2010).

3.8.1. Performance characteristics of qPCR

Performance characteristics which include efficiency, limit of detection and sensitivity of amplicons were determined by amplifying three separately prepared sets of dilution series of three standard samples in water which include1- phytoplasma-infected periwinkle (phytoplasmal DNA) 2- phytoplasma-infected sugarcane (phytoplasmal DNA) 3- phytoplasma-free sugarcane (plant DNA). Since the copy number of target genes in the standard samples is unknown, the standard curves are helpful for the evaluation of PCR efficiency and sensitivity, but not for absolute quantification.

3.8.1.1. Efficiency Measurement

In this study, efficiency (E) values were measured using the Ct slope method. This method involves generating a dilution series of the target template and determining the Ct value for each dilution. A plot of Ct versus log DNA concentration is constructed(Figures.3.29, 30 and 31). Amplification efficiency was calculated from the slope of this graph using the equation: $Ex = 10^{(-1/slope)} - 1$. The effect of efficiency is exponentially dependent on cycle number. If E=1, amplicon quantity is duplicated every cycle. If E=0.8 amplicon quantity is only duplicated every 1. 2 cycle. The squared regression coefficient after the linear regression (R^2) was also determined (Table.3.16 and 17).

Results

Standard curve of qPCR

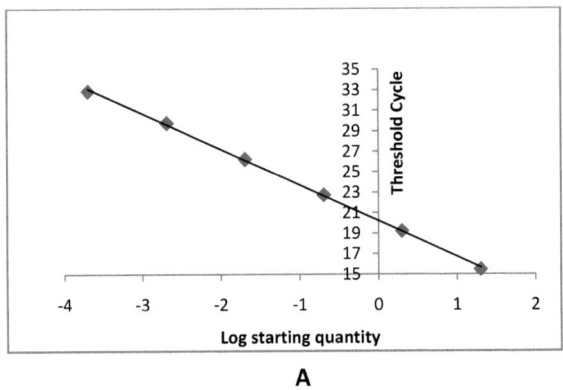

A

Amplification chart

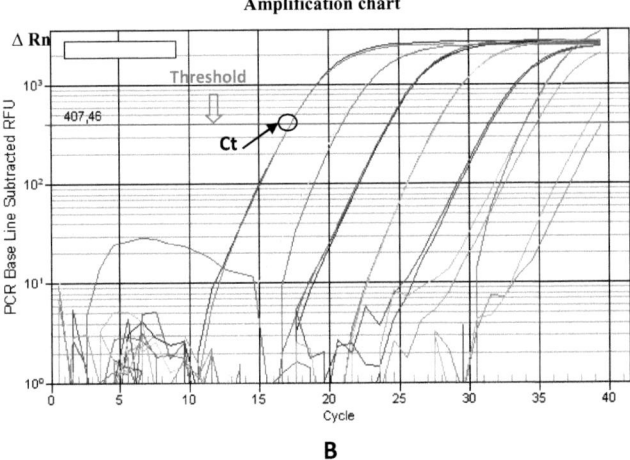

B

Figure.3.29. A:Standard curve. Standard curve determined at six concentration levels (ranging from $10^{\wedge}0$ to $10^{\wedge}-5$) using 10-fold dilution series of the reference sample (phytoplasma-infected periwinkle). The threshold numbers of PCR cycles (CT value; means of triplicates) are plotted against the dilution (log scale). **B: Log- view of standard curve chart. Threshold;** is an arbitrary level of fluorescence chosen on the basis of the baseline variability. **Ct;** is defined as the fractional PCR cycle number at which the reporter fluorescence is greater than the threshold. ΔRun; is an increment of fluorescent signal at each time point. The ΔRun values are plotted versus the cycle number.

Results

Standard curve of qPCR

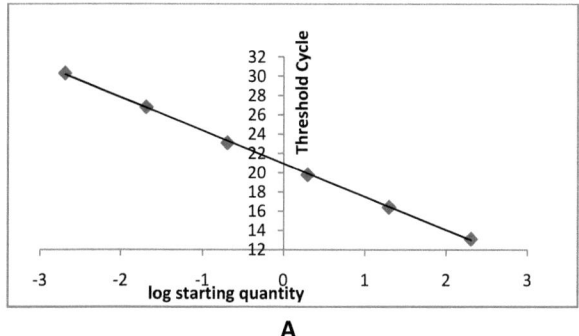

A

Amplification chart

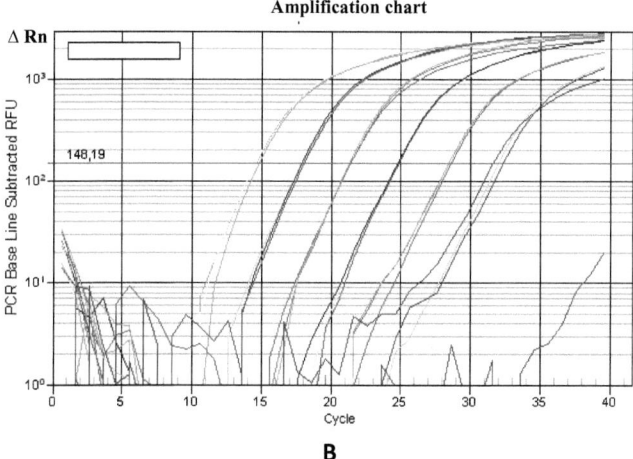

B

Figure.3.30. A:Standard curve. Standard curve determined at six concentration levels (ranging from 10^0 to 10^-5) using 10-fold dilution series of the reference sample (phytoplasma-infected sugarcane). The threshold numbers of PCR cycles (CT value; means of triplicates) are plotted against the dilution (log scale). **B: Log- view of standard curve chart.**ΔRun; is an increment of fluorescent signal at each time point. The ΔRun values are plotted versus the cycle number

Results

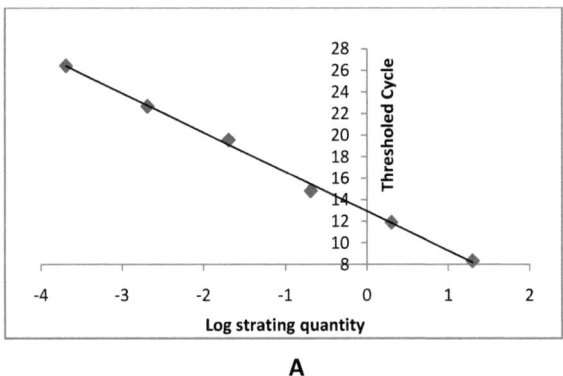

A

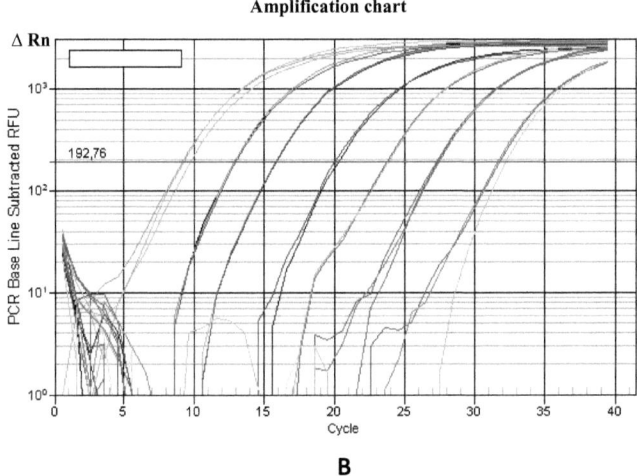

B

Figure.3.31. A:Standard curve. Standard curve determined at six concentration levels (ranging from 10^0 to 10^{-5}) using 10-fold dilution series of the phytoplasma-free sugarcane plant sample. The threshold numbers of PCR cycles (CT value; means of triplicates) are plotted against the dilution (log scale). **B: Log- view of standard curve chart.** ΔRun; is an increment of fluorescent signal at each time point. The ΔRun values are plotted versus the cycle number

3.8.1.2. Artificial samples to test sensitivity of qPCR assay

Performance characteristics were also evaluated for the serial dilution of phytoplasma-infected sugarcane (phytoplasmal DNA) mixed with sugarcane DNA, instead of water, isolated from phytoplasma-free sugarcane leaves to imitate real infected sugarcane samples. Thus, PCR sensitivity was evaluated for potential effects of host-material inhibition. This un-infected sugarcane material had already been tested and confirmed to be phytoplasma-free sugarcane. Artificial samples imitating infected sugarcane samples were prepared by serial dilutions of phytoplasma-infected sugarcane DNA mixed with phytoplasma-free sugarcane DNA. A plot of Ct versus log DNA concentration is also constructed as above (Figure.3.32).

Results

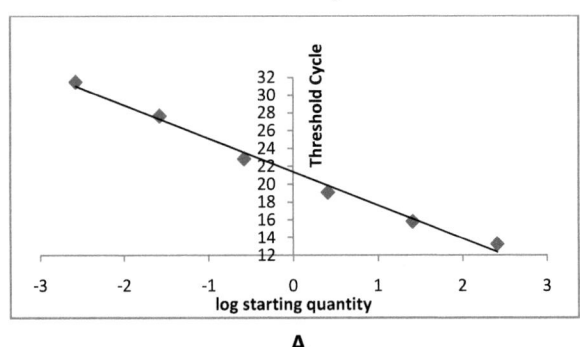

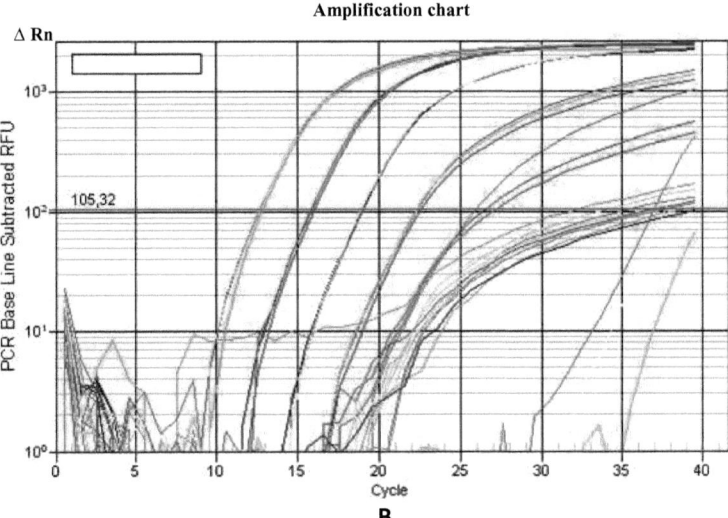

Figure.3.32. A: Standard curve. Standard curve determined at six concentration levels (ranging from 10^0 to 10^{-5}) using 10-fold dilution series of the phytoplasma-infected sugarcane (phytoplasmal DNA) mixed with sugarcane DNA to imitate real sugarcane samples. The threshold cycle numbers (CTvalue; means of triplicates) are plotted against the dilution (log scale). **B: Log-view of standard curve chart.** ΔRun; is an increment of fluorescent signal at each time point. The ΔRun values are plotted versus the cycle number.

Results

Table.3.16. Performance characteristics of qPCR assay. Performance characteristics measured for three standard samples used in this study in water dilutions which include phytoplasma-infected periwinkle (phytoplasmal DNA), phytoplasma-infected sugarcane (phytoplasmal DNA) and phytoplasma-free sugarcane (sugarcane plant DNA) and measured also for one standard sample which is phytoplasma-infected sugarcane (phytoplasmal DNA) but in sugarcane plant DNA dilutions (approximately 200 ng of genomic DNA per reaction in undiluted samples).

target template	nocilpma	epols	R squared	fo egnar noitceted	cimanyd egnar	PCR efficiency (%)	fo timil noitceted
phytoplasma infected periwinkle	16S rRNA (retaw)	-3.5008	0.9994	$1\text{-}10^{-5}$	$1\text{-}10^{-5}$	90.5%	32.90
phytoplasma infected sugarcane	16S rRNA (retaw)	-3.4483	0.9996	$1\text{-}10^{-5}$	$1\text{-}10^{-5}$	94.9%	30.31
phytoplasma infected sugarcane	16S rRNA (tnalp AND)	-3.7381	0.9905	$1\text{-}10^{-5}$	$1\text{-}10^{-4}$	86 %	26.54
phytoplasma free sugarcane	18S rRNA (retaw)	-3.647	0.9971	$1\text{-}10^{-5}$	$1\text{-}10^{-5}$	87%	26.41

Table.3.17. Q-PCR results from 10-fold dilution series of the reference samples which diluted in water or in healthy host plant DNA. CT is a threshold cycle number of qPCR assay.

reference sample	type of dilution	CT mean value 10^0 dilution	CT mean value 10^{-1} dilution	CT mean value 10^{-2} dilution	CT mean value 10^{-3} dilution	CT mean value 10^{-4} dilution	CT mean value 10^{-5} dilution
phytoplasma infected periwinkle	retaw	15.49	19.22	22.75	26.27	29.79	32.90
phytoplasma infected sugarcane	retaw	13.14	16.40	19.78	23.11	26.80	30.31
phytoplasma infected sugarcane	host plant DNA	12.72	15.86	18.95	22.39	26.54	33.96
phytoplasma free sugarcane	retaw	08.35	11.92	14.82	19.59	22.68	26.41

3.8.2. Q-PCR results of the sugarcane samples from different sources

The legend for all next tables is as the following: CT is a threshold cycle number of qPCR assay. ND, phytoplasma not detected, i.e. above the maximum CT value of 40 cycles.

Table.3.18.Q-PCR results of the Hawaiian sugarcane samples grown in greenhouse.

Sample	CT mean value (phytoplasma 16S assay)	CT mean value (plant 18S assay)
H65-7052	34.93	11.95
H78-7750	35.30	12.82
H87-4094	ND	11.01
H87-4319	ND	11.75
Positive control	14.93	10.04
Water control	ND	ND

Table.3.19.Q-PCR results of the Egyptian and Syrian sugarcane samples.

Sample	CT mean value (phytoplasma 16S assay)	CT mean value (plant 18S assay)
Gt549	ND	10.42
G8447	ND	09.96
Syrian cultivar	ND	11.55
Positive control	14.65	11.47
Water control	ND	ND

Table.3.20. Q-PCR results of the Hawaiian plantations sugarcane samples (2009). The samples (K1-2 to K1-5) were collected from G&R plantation in Kauai (4-5 km west of Olokele) while the samples (H1 and H3) were from former plantations in Hawaii and the sample (M11) was from former plantation in Maui.

Sample	CT mean value (phytoplasma 16S assay)	CT mean value (plant 18S assay)
K1-2	32.25	09.37
K1-3	ND	08.95
K1-4	ND	09.93
K1-5	29.77	10.98
H1	ND	10.26
H3	ND	09.17
M11	ND	13.34
Positive control	14.93	10.04
Water control	ND	ND

Results

Table.3.21. Q-PCR results of the Hawaiian breeding station sugarcane samples (2010).

Sample	CT mean value (phytoplasma 16S assay)	CT mean value (plant 18S assay)
7052	ND	08.28
6110	ND	08.64
4094	ND	08.98
Positive control	14.93	10.04
Water control	ND	ND

Table.3.22. Q-PCR results of the Hawaiian grass weeds from Sumida watercress farm (2011).

Sample	CT mean value (phytoplasma 16S assay)	CT mean value (plant 18S assay)
Paspalum coniugatum	ND	08.54
Cyperus rotundus	ND	10.68
Panicum maximum	ND	08.26
Miscanthus Sp.	ND	10.42
Positive control	13.34	08.88
Water control	ND	ND

Table.3.23. Q-PCR results of the Hawaiian sugarcane samples from HC&S plantation in Maui island, close to Puunene (2011).

Sample	CT mean value (phytoplasma 16S assay)	CT mean value (plant 18S assay)
H65-7052, 6 months Non-symptomatic	ND	09.04
H65-7052, 6 months symptomatic	ND	10.53
H73-3567, 4 months Non-symptomatic	33.19	08.15
H87-4319, 9 months Non-symptomatic	34.51	07.86
H87-4319, 9 months slightly-symptomatic	ND	11.24
H86-3792, 6 months Non-symptomatic	33.10	10.74
H87-5794, 9 months Non-symptomatic	32.83	08.14
Positive control	11.95	08.21
Water control	ND	ND

Table.3.24. Q-PCR results of the Hawaiian sugarcane samples from HARC in Maunawili (2011).

Sample	CT mean value (phytoplasma 16S assay)	CT mean value (plant 18S assay)
H87-4094 field A11	ND	10.05
H78-3567 "	ND	10.20
H87-4319 "	36.75	08.96
H65-7052 "	ND	08.60
H50-7209 "	ND	08.80
H78-4153 "	37.47	09.13
H77-4643 "	37.29	08.60
H73-6110, field A22	ND	09.67
H32-8560, field B31		
H78-7750, field B62	ND	09.90
H87-4319 "	ND	09.31
H78-3606 "	ND	10.34
H77-4643, field P11a	ND	08.92
H78-3606, field P11	ND	10.84
H50-7209, field P12	ND	09.38
H87-4319 "	ND	10.75
H65-7052 "	ND	09.12
H78-7750 "	36.42	07.96
H73-6110, field P13	33.98	11.59
Positive control	11.43	08.21
Water control	ND	ND

Table.3.25. Q-PCR results of the Hawaiian plantations sugarcane samples from different sources close to former plantations (2011).

Sample	CT mean value (phytoplasma 16S assay)	CT mean value (plant 18S assay)
Virus-free from Dr. Lehrer	ND	08.55
Honomu, Stable Camp Rd	ND	09.31
Honomu, Akaka Falls Rd	ND	09.06
Honomu, Akaka Falls Rd	ND	08.89
Akaka Falls, 300m elevation	ND	09.00
" , 200m elevation	ND	08.82
Kukui Camp	ND	09.58
Keanae, Hana Rd., close to Taro	ND	08.59
Positive control	12.87	08.66
Water control	ND	ND

Table.3.26. Q-PCR results of the Thai sugarcane samples from Bang Phra and Khon Kean provinces (2010).
Samples (F1 to F8) were from farmer fields in province of Bang Phra while the samples (S1 to S13) were from breeding station also in province of Bang Phra. Samples (KK1 to KK38) were from province of Khon Kean.

Sample	CT mean value (phytoplasma 16S assay)	CT mean value (plant 18S assay)
F1	23.86	10.50
F2	29.10	10.02
F3	24.11	13.66
F4	17.36	11.12
F5	17.98	12.30
F6	ND	10.97
F7	ND	12.07
F8	17.34	13.28
S1	ND	10.46
S2	ND	10.84
S3	ND	11.55
S4	ND	10.70
S5	ND	11.96
S10	ND	11.79
KK1	17.70	10.05
KK2	ND	12.42
KK3	18.78	10.50
KK4	21.04	12.20
KK11	31.98	12.87
KK12	ND	08.75
KK13	ND	08.09
KK14	25.58	11.27
KK17	ND	11.05
KK18	ND	11.41
KK20	ND	12.37
KK32	26.87	11.70
KK34	17.74	11.17
KK35	ND	11.08
KK36	ND	10.82
KK37	ND	10.63
KK38	34.83	12.18
AAY (positive control)	17.34	14.05
Water control	ND	ND

Table.3.27. Q-PCR results of the Thai sugarcane samples from Suphan Buri province.

Sample	CT mean value (phytoplasma 16S assay)	CT mean value (plant 18S assay)
1	ND	10.33
2	ND	10.36
3	ND	14.27
4	ND	10.89
Grassy shoot	16.98	07.82
positive control	14.65	12.24
Water control	ND	ND

Results

The Ct values differed considerably among samples in the phytoplasma assay, while Ct values obtained in the plant assay were different slightly. This result indicated that phytoplasma titer was variable.

3.8.3. Distribution of phytoplasma in sugarcane

Q-PCR assay of phytoplasma 16S DNA was used to determine the distribution of the phytoplasma within infected sugarcane plant. The relative distribution of sugarcane white leaf phytoplasma in different parts of the plant was quantified using the comparative Ct method. Three leaf samples of phytoplasma infected sugarcane including white, variegated and green; and root samples were analysed (Figure.3.33). The phytoplasma was detected in all tested organs including leaves and roots. It seems there is correlation between titer of phytoplasma and symptoms expression where the titer of phytoplasma in white leaf was higher than variegated and green leaves. Lower Ct values correspond to higher initial quantities of phytoplasma DNA template (Table.3.28).

Table.3.28. Q-PCR results of the distribution of the phytoplasma in sugarcane plant. *CT* is a threshold cycle number of qPCR assay. +,phytoplasma detected.

Sugarcane sample	*CT* mean value (16S) Phytoplasma	*CT* mean value (18S) Sugarcane	Phytoplasma detection
White leaf	11.56	08.31	+
Variegated leaf	12.97	08.76	+
Green leaf	14.02	08.79	+
Root	12.28	08.82	+

Results

Figure.3.33. A: Thai sugarcane plant infected with sugarcane white leaf phytoplasma where some leaves are totally bleaching whereas others are variegated and some green leaves also exists. This picture was taken three months post germination comparison with non-infected sugarcane plants in **(B)**.

3.9. Phylogenetic analysisof the phytoplasma strains in sugarcane

Restriction fragment analysis had been successfully applied to differentiate between the phytoplasma strains (Kirkpatrick et al., 1994; Lee et al., 1998; Valiunas et al., 2007). The products of the second round PCR were subjected to restriction enzymes which were diagnostic for the phytoplasma strains. The restriction patterns identified the phytoplasma from Hawaiian cultivars and from one Cuban cultivar as belonging to the Aster yellows phytoplasma, whereas the phytoplasma from the Cuban cultivar CP4362, which originally had been bred in Canal Point, Florida and from JA605, belonged to the Western X-disease phytoplasma (Figure.3.4 and Table. 3.2). This classification was supported by sequence comparison.

The complete sequence of R16F2n/R16R2-amplified fragments was determined for three different sugarcane cultivars which are infected by three different phytoplasma isolates, two cultivars are from Cuba and one from Egypt (Figure.3.34). The complete sequence of 16S/23S intergenic spacer region was determined for one Hawaiian sugarcane cultivar using the primer pair P4/P7 (Figure.3.34). The partial sequence of 16S/23S intergenic spacer region was also determined for other two Hawaiian sugarcane cultivar and for one cultivar from Thailand using the primer pair P1/P2 (Figure.3.34).

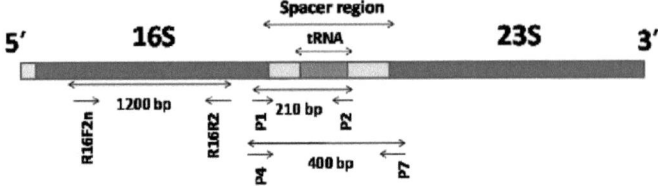

Figure.3.34. Diagrammatic representation of genomic location of primers used for DNA sequencing.

The obtained nucleotide sequences were compared with sequences of phytoplasmas and acholeplasmas from GenBank using the BLASTN program. Multiple alignments of near-full-length 16S rRNA gene sequences from 22 phytoplasma and one Acholeplasma species and multiple alignments of 16S/23S intergenic spacer region from 13 phytoplasma and two Acholeplasma species were examined using MUSCLE software. Phylogenetic trees of both sequence parts were constructed to reveal the position of the isolated phytoplasma strains from Hawaiian, Cuban, Egyptian and Thai sugarcane, relative to phytoplasma strains which had been isolated from sugarcane and other plants. Figure.3.35presentsthe two phylogenetic trees that were constructed by maximum likelihood estimation with geneious program through the PhyML software (Guindon and Gascuel, 2003). Bootstrap analysis was performed 1.000 times to evaluate branch supports in a sound statistical framework.

The phytoplasma isolate (HM804282) from Cuban sugarcane cultivar Ja605 clustered together with other strains of X-disease group, among them already reported sugarcane yellows phytoplasma strain found in South Africa (AF056095) with a bootstrap value of 48.7and shared 99% sequence identity (Figure.3.35 andTable.3.29). Other Cuban sugarcane cultivar C10-5173 was infected with phytoplasma strain (HQ116553) clustered to the aster yellows group, closely together with sugarcane yellows phytoplasma from Brazil (EU423900) and maize bushy stunt phytoplasma from Colombia (HQ530152) with a bootstrap value of 49.6 and shared 99% sequence identity (Figure.3.35and Table.3.29).

Egyptian sugarcane cultivar G8447 contains phytoplasma strain (JN223446) clustered to the rice yellow dwarf group, closely together with sorghum grassy shoot phytoplasma from Australia (AF509324) with a bootstrap value of 81.1 and shared 99% sequence identity (Figure.3.35and Table.3.29).

Results

The phylogenetic tree of the 16S/23S spacer region contained less phytoplasma entries in GenBank. The Hawaiian sugarcane phytoplasma isolate(HQ116554) from cultivar H84-4094 and another Hawaiian sugarcane phytoplasma isolate(JN223447) from unknown cultivar obtained from Hawaiian former plantations as a different original source clustered to the aster yellows group, closely together with water cress yellows from Hawaii (AY665676) and Russian potato purple top phytoplasma (EU333399) with a bootstrap value of 69.6and shared 99% sequence identity (Figure.3.35and Table.3.29).

Hawaiian sugarcane phytoplasma isolate(JN223448)from cultivar H78-7750which obtained from Hawaiian breeding station of HARC in Maunawili, Oahu clustered to the rice yellow dwarf group, closely together with sugarcane white leaf phytoplasma from Taiwan (AY139874) with a bootstrap value of 86.6and shared 98% sequence identity (Figure.3.35andTable.3.29).

Thai sugarcane phytoplasma isolate(HQ917068) from unknown sugarcane cultivar obtained from province of Khon kaen clustered to the rice yellow dwarf group, closely together with sugarcane white leaf phytoplasma from Myanmar (AB646271) with a bootstrap value of 64.2and shared 100% sequence identity (Figure.3.35and Table.3.29).

Table.3.29.Phytoplasma strains and their GenBank accession numbers used in this study for the phylogenetic trees (Figure.3.35). 16S rRNA gene and 16S/23S intergenic spacer region sequences of phytoplasmas determined in this study are in bold. Phytoplasma strains of monocotyledonous plants and strains which showed close sequence similarity to the Hawaiian, Cuban, Egyptian and Thai sugarcane phytoplasma were selected for construction of the trees. The sequences from *Acholeplasma axanthum* and *Acholeplasma palmae* were used as out groups.

Phylogenetic tree of 16S rRNA (a)		
Accession number	Phytoplasma strain	Group
AF056095	Sugarcane yellows phytoplasma type I (South Africa)	X-disease
AF411592	Erigeron witches'-broom phytoplasma	Ash yellows
AF509324	Sorghum grassy shoot phytoplasma variant I (Australia)	Rice Yellow Dwarf
AF498307	Coconut lethal yellowing phytoplasma	Coconut lethal yellowing
AJ550984	Bermuda grass white leaf phytoplasma (Southern Italy)	Bermuda white leaf
AM261831	Sugarcane grassy shoot phytoplasma (India)	Rice Yellow Dwarf
AY197652	Spartium witches'-broom phytoplasma	Apple proliferation
AY736374	Napier grass stunt phytoplasma (Kenya)	Rice Yellow Dwarf
EF413055	Sorghum verticilliflorum phytoplasma (Mauritius)	X-disease
EF413056	Sugarcane yellows phytoplasma clone SC245 (Mauritius)	X-disease
EU294011	Malaysia Bermuda grass white leaf phytoplasma	Bermuda white leaf
EU423900	Sugarcane yellows phytoplasma type I (Brazil)	Aster yellows
FM208260	Sugarcane white leaf (Thailand)	Rice Yellow Dwarf
GQ336993	Kidney bean little leaf phytoplasma clone Z16 16S	Peanut WB
GQ850122	Coconut root wilt phytoplasma isolate RD3 (India)	Rice Yellow Dwarf
GU565959	Candidatus Phytoplasma pyri isolate 932801	Apple proliferation
HM804282	**Sugarcane Ja60-5 yellow leaf (Cuba)**	**X-disease**
HQ116553	**SugarcaneC1051-73 yellow leaf (Cuba)**	**Aster yellows**
HQ530152	Maize bushy stunt phytoplasma strain MBSColombia (Colombia)	Aster yellows
HQ589200	Milkweed yellows phytoplasma strain MWI(USA)	X-disease
JF508514	Sesame phyllody phytoplasma strain Seph2	Peanut WB
JN223446	**Sugarcane grassy shoot phytoplasma (Egypt)**	**Rice Yellow Dwarf**
NR_029152	Acholeplasma palmae strain J233	Out group

Results

Phylogenetic tree of 16S/23S intergenic spacer region(b)		
Accession number	Phytoplasma strain	Group
AB243298	Sugarcane grassy shoot phytoplasma (India)	Rice Yellow Dwarf
AB646271	Sugarcane white leaf phytoplasma (Myanmar)	Rice Yellow Dwarf
AF434989	Texas Phoenix palm phytoplasma	Coconut lethal yellowing
AY139874	Sugarcane white leaf phytoplasma (Taiwan)	Rice Yellow Dwarf
AY665676	Aster yellows phytoplasma "Watercress" (Hawaii)	Aster yellows
DQ004923	Acholeplasma palmae	Out group
DQ400425	Acholeplasma axanthum	Out group
EU294011	Malaysia Bermuda grass white leaf phytoplasma	Bermuda white leaf
EU333399	Russian potato purple top phytoplasma (Russia)	Aster yellows
FN562932	Candidatus Phytoplasma vitis	Elm yellows
HQ116554	**Hawaiian sugarcane H87-4094 yellow leaf phytoplasma**	**Aster yellows**
HQ589192	'Psammotettix cephalotes' flower stunt phytoplasma	Rice Yellow Dwarf
HQ917068	**Sugarcane white leaf phytoplasma (Thailand)**	**Rice Yellow Dwarf**
JN223447	Hawaiian sugarcane Phytoplasma	Aster yellows
JN223448	Sugarcane white leaf phytoplasma (Hawaii)	Rice Yellow Dwarf

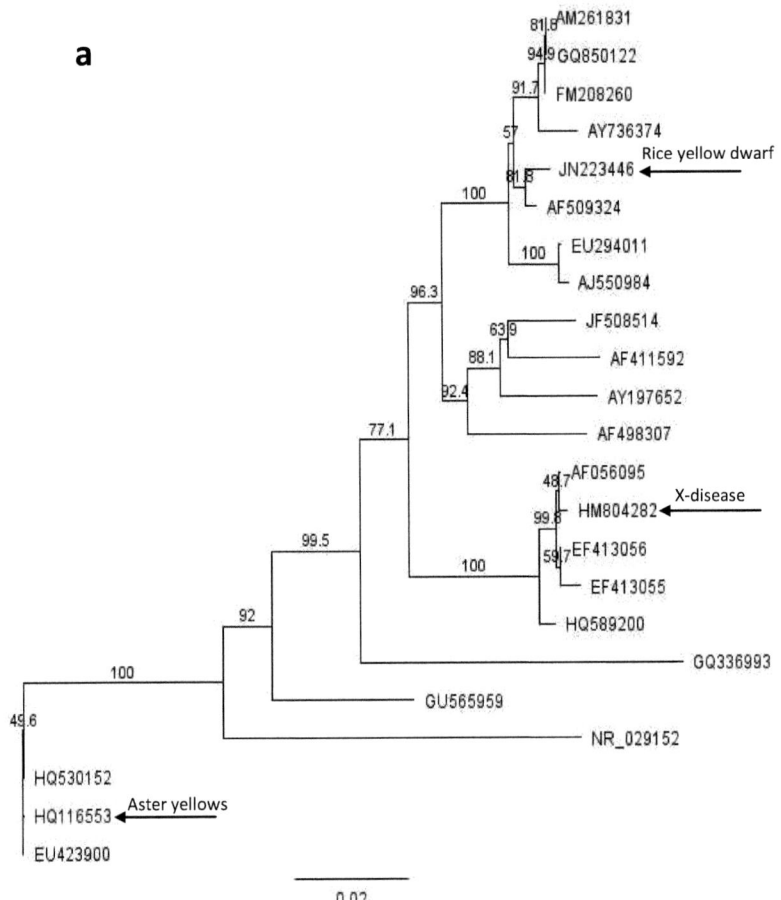

Results

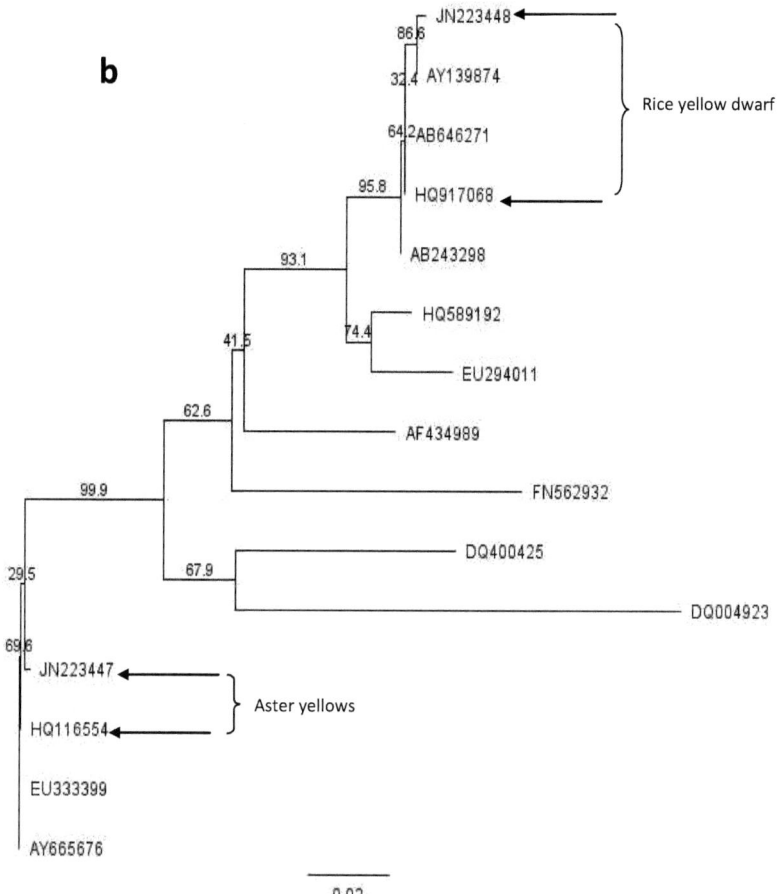

Figure.3.35.Position of the phytoplasma strains from Hawaiian, Cuban, Egyptian and Thai sugarcane in a phylogenetic tree together with other phytoplasma isolates(Table.3.29).a: Phylogenetic tree constructed using 16S rRNA sequences from 22 phytoplasma and one Acholeplasma species, b: Phylogenetic tree constructed using 16S/23S spacer sequences from 13 phytoplasma and two Acholeplasma species. Bar represents phylogenetic distance of 2%. Numbers on branches are confidence percentage obtained from 1.000 bootstrap replicates.

3.10. Hot water treatment in order to get phytoplasma free sugarcane plant

Hot water treatment had been proposed as a cure for phytoplasma in dormant woody plant material (Edison and Ramakrishnan, 1972; Caudwell et al., 1997), because phytoplasmas have only limited heat tolerance.

Treatment of infected sugarcane stalks with moderately high temperatures such as 50°C for 2 h were reported to successfully eliminate grassy shoot disease and white leaf phytoplasma from stem material. Hot water treatment of stem cuttings together with immersion in a fungicide solution is a routine practice in Hawaiian sugarcane plantations to prevent fungal rot of planted seed pieces. The question was which temperature regimes and which incubation durations are needed to eliminate SCYLP from sugarcane stems and whether the routine hot water-treatment against fungi had unintentionally also cured from phytoplasma.

One-eye stem cuttings were immersed in hot water of defined temperature and for defined period, then planted in sterile soil in pots and kept in insect-tight mesh cages for germination and growth. Indeed, incubation of seed pieces at 50°C for 30 min or longer was sufficient to eliminate phytoplasma (Table.3.30), irrespective whether it was from Aster yellows or from Western X-disease type. The incubation in hot water for 3 h had a detrimental effect on seed piece viability unless the hot water treatment was preceded by 10°C incubation for 48h, a procedure routinely used in the Australian and Cuban sugar industry.

3.10.1. Hot water treatment according to Australian recipe

Two sugarcane cultivars (H65-70 52 and H78-77 50) were used as material in the hot water treatments. Stalks were cut into single-eye sets and treated by immersion for 48 h in cold water (10°C) followed by 3h in hot water (50°C). Next, setts were planted in sterile soil and placed in mesh cage to protect them against insects. Subsequently, these plants which rose from these cuttings were tested for the presence of phytoplasma after 2 and 6 months and one year post germination (Figure.3.36).

Results

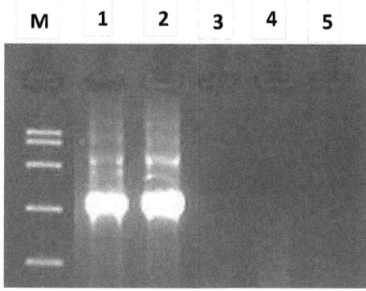

Figure.3.36. Nested PCR results of the test plants, which received cold and hot-water treatment after 2 months post germination. Lane (1) contains PCR product obtained from untreated sugarcane (without hot water treatment) (H65- 70 52). Lane (2) contains PCR product obtained from untreated sugarcane (H78- 77 50). Lane (3) test plant (H65- 70 52) after cold- and hot-water treatment (Australian recipe). Lane (4) test plant (H78- 77 50) after cold- and hot-water treatment (Australian recipe). Re-amplification of aliquot of first PCR water control with nested primer combination is in lane (5). DNA ladder is FastRuler Middle Range (MBI Fermentas).

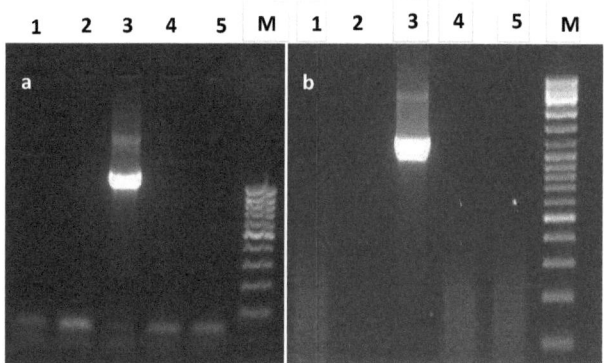

Figure.3.37. Nested-PCR results of test plants, which received cold and hot-water treatment after 6 months (a) and 1 year (b) post germination. Lane (1) treated plant (H65-70 52); lane (2) treated plant (H78-77 50); lane (3) is positive control; re-amplification of aliquots of first PCR water control with nested primer combination are in lanes (4-5); lane (M) GeneRuler DNA ladder (MBI Fermentas).

3.10.2. Hot water treatment with various duration

These tests were carried out in order to investigate what is the minimum immersion time at 50°C can eliminate the phytoplasma in infected sugarcane plants. Furthermore, in these tests we didn't soak the cuttings in cold water at 10°C before hot water treatment in order to check the influence of lacking of cold water treatment on the vegetative development. Thus, the cuttings from cultivar (H78-77 50) were treated by immersion directly in hot water at 50° C for 30 min, 1h, 2h and 3h. For each treatment, three replicates were tested (Figure.3.38 and Figure.3.39).

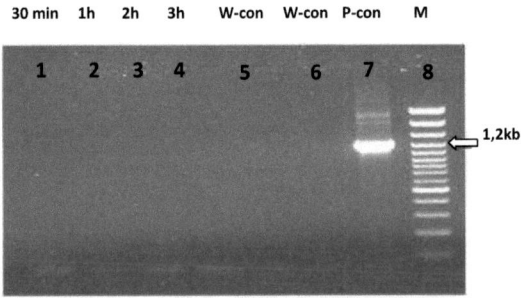

Figure.3.38. Nested-PCR results of test plants; which received hot water treatment at 50°C with various duration after 2 months post germination. Lanes (1-4) treatedplants (H78-77 50) with various duration; re-amplification of aliquots of first PCR water controls with nested primer combinations are in lanes (5-6); lane (7) positive control. Lane 8, GeneRuler 100pb plus DNA ladder (MBI Fermentans).

Note: tested plants which were treated at 50°C for 3h had a very poor vegetative development and later died.

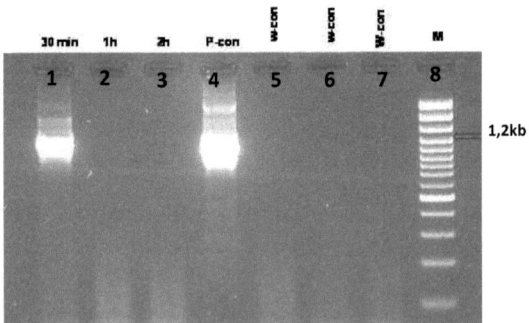

Figure.3.39. Nested-PCR results of test plants; which received hot water treatment at 50°C with various duration after 6 months post germination. lanes (1-3) test plants (H78-77 50) with various duration (30 min,1h, 2h); lane (4) positive control; re-amplification of aliquots of first PCR water controls with nested primer combination are in lanes (5-6-7); lane (8) GeneRuler 100pb plus DNA ladder (MBI Fermentans).

Table.3.30. **Effect of hot water treatment on phytoplasma elimination of cultivars H65-7052 and H78-77 50 according to the indicated temperature regime.** The data are from 3 replicates of each treatment.

Treatment	Time of testing after treatment	Presence of phytoplasma
50°C, 30 min	2 months and 6 months	+
50°C, 1 h	2 months and 6 months	-
50°C, 2 h	2 months and 6 months	-
50°C, 3 h	2 months, no viable plants after 6 months	-
48 h 10°C, then 3 h 50°C	2 months, 6 months and 1 year	-

3.11. Insecticide treatment of phytoplasma-infected sugarcane plant

We wanted to test if the insecticide treatment of phytoplasma-infected sugarcane plant may reduce the titer of phytoplasma or not. Therefore, three white leaf phytoplasma-infected sugarcane plants were treated with Confidor as a systematic insecticide. One untreated plant was used as control. Q-PCR assay of phytoplasma 16S rRNA gene was used to determine the phytoplasma infection level within infected sugarcane plants. The relative concentration of sugarcane white leaf phytoplasma in three different treated plants and one non-treated plant was quantified three months post treatment using the comparative Ct method. Based on Ct mean values of qPCR, the insecticide treatment has no effects on the phytoplasma infection level (Table.3.31).

Table.3.31. **Q-PCR results of the insecticide treatment of phytoplasma-infected sugarcane plants.** Three sugarcane white leaf phytoplasma-infected plants were treated with Confidor as a systematic insecticide. One untreated plant was used as control. The relative quantification was carried out three months post-treatment and Ct mean values of qPCR revealed that the insecticide treatment can not reduce the titer of phytoplasma.

Sugarcane sample	Ct mean value (16S) Pre-treatment	Ct mean value (16S) Post-treatment	Ct mean value (18S) Pre-treatment	Ct mean value (18S) post-treatment
Insecticide treated plant 1	12.98	12.84	07.83	07.71
Insecticide treated plant 2	12.13	12.09	07.70	07.91
Insecticide treated plant 3	11.32	12.23	07.97	07.52
Untreated plant (control)	12.06	12.27	08.38	08.31

3.12. Transmission test with sugarcane aphid (*Melanaphis sacchari*)

Phytoplasmas are phloem-limited; therefore, only phloem-feeding insects can potentially acquire and transmit the pathogen. Thus far, however, there has been no report of phytoplasma or spiroplasma transmission by a phloem-feeding aphid; the reasons for lack of transmission by aphids are not known (Mishra, 2004).

Aphids (*Melanaphis sacchari*) were collected in Hawaii and were brought to our lab as alive insects and then were established in a mesh cage on sugarcane plants in a climate chamber. Phytoplasma-infected sugarcane plants were transferred to a cage together with the aphids. After 45 days of acquisition and latency period, phytoplasma–free plants were transferred into the same cage for inoculation feeding and these plants were kept there for 3 months (Figure.3.40). Our tests showed that these aphids are able to acquire the phytoplasmas but they do not serve as phytoplasma vectors.

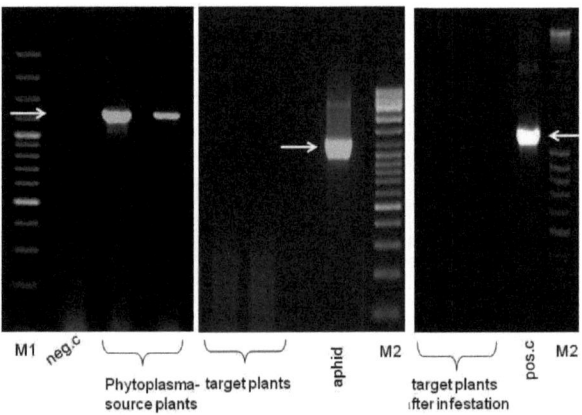

Figure.3.40. Phytoplasma in sugarcane source plants (containing phytoplasma), in aphids (*Melanaphis sacchari*) and in sugarcane target plants (infested by phytoplasma-infected *M. sacchari*). The phytoplasma-infected plants (cv. H65-7052) were infested with aphids for 45 days (acquisition-feeding). The aphids were tested for phytoplasma and then transferred to target plants (cvs. H65-7052 and H78-7750). The target plants had previously been made phytoplasma-free by cold and hot water treatment. The target plants were tested 3 months after inoculation with phytoplasma-infected aphids. The negative control is the phytoplasma-free cv. Ph-8013. The marker M1 was GeneRuler 100 bp plus (MBI Fermentas), the marker M2 was GeneRuler DNA ladder Mix (MBI Fermentas) on right gel. The arrows point to the phytoplasma-specific band of 1.2 kbp.

3.13. Transmission electron microscopy for cytological location of phytoplasma

For many phytoplasma species it is known that the infection is mainly located in sieve elements of phloem cells, to which phytoplasmas are introduced by phloem-feeding homopteran insects, mainly leafhoppers and planthoppers (Weintraub and Beanland, 2006). Using transmission electron microscopy we want to show if also for sugarcane white leaf phytoplasma "*Ca.*Phytoplasma oryzae" the infection is mainly restricted to phloem sieve elements or if other cell types are infected as well. Since, a few phloem parenchyma cells adjacent to sieve tubes are occasionally also invaded by phytoplasma (Christensen et al., 2005). For sugarcane known for the high cytoplasmic sucrose contents this could be also the case. Secondly we wanted to study the ultrastructural effects of the infection of phytoplasma.

3.13.1. Anatomy of leaf phloem tissue

The vascular tissue including xylem and phloem are found within the veins of the leaf.
Phloem cells are usually located next to the xylem which made mainly from vessel elements and parenchyma cells. The two most common cells in the phloem are the companion cells and sieve tube cells; (Figure.3.41).

Phloem is made from columns of parenchyma cells. Each parenchyma cell is adapted to form a sieve element .Columns of sieve elements join together to form sieve tubes. The cross walls between successive cells (sieve elements) become perforated forming sieve plates. As the sieve elements mature they loose several plant cell organelles – the nucleus, ribosomes and Golgi body degenerate. This allows materials to pass through them more easily. Sieve elements have a thin cell wall, cell membrane, and may be plastids and the lumen is filled with sap.

Each sieve element has at least one companion cell next to it. Companion cells have the normal plant cell structure with extra ribosomes and mitochondria. Companion cells are linked to the sieve elements by numerous plasmodesmata. As might be expected, it is companion cells that enable the sieve element to stay alive.

The outer layer of the vein is made of cells called bundle sheath cells, and they create a circle around the xylem and the phloem. They form a protective covering on leaf veins, and consist of one or more cell layers, usually parenchyma. Loosely arranged mesophyll cells lay between the bundle sheath and the leaf surface;(Figure.3.41). Bundle sheath and mesophyll cells are packed with chloroplasts, and this is where the dark reactions of photosynthesis are actually occurring. The air spaces between the cells allow for gas exchange.

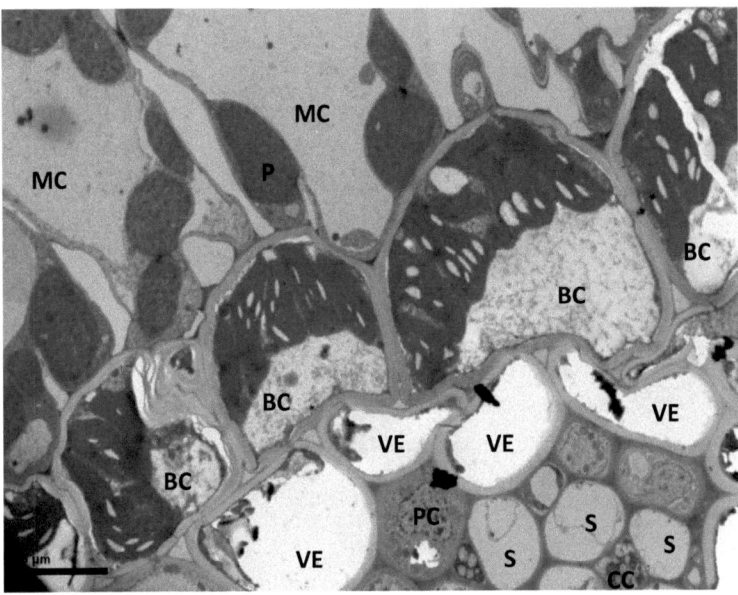

Figure.3.41. Ultrathin section of green leaf of uninfected sugarcane (control). This ultrathin section shows structure of phloem, xylem, bundle sheath, and mesophyll cells. Phloem is made from sieve cells (S), companion cells (CC), and parenchyma cells (PC), while vessel elements (VE) are the main components of xylem tissue. The outer layer of the vein is made of cells called bundle sheath cells (BC), and they create a circle around the xylem and the phloem. Loosely arranged mesophyll cells (MC) lay between the bundle sheath and the leaf surface. Bundle sheath and mesophyll cells are packed with plastids (P). Bar = 5 µm.

3.13.2. Localization of phytoplasma infection

Our transmission electron microscopic studies revealed the presence of sugarcane white leaf phytoplasma "*Ca.* Phytoplasma oryzae" of regular shape and size in phloem sieve tubes of diseased sugarcane leaves including white; (Figure.3.42.A), variegated; (Figure.3.42.B), and green leaves; (Figure.3.42.C). Phytoplasmas were observed in mature and immature phloem sieve tubes. However, no sugarcane white leaf phytoplasmas are present in green leaf of uninfected sugarcane; (Figure.3.42.D), where the vacuoles and cytoplasm are fused to a so called mictoplasm (Esau et al., 1965). Most of the organelles are absent, only typical round shaped sieve elements plastids are present. Sieve element plastids show the typical phenotype with christae inclusions which are common for poaeceae (arrow Figure.3.42.D). Surrounding companion cells are phytoplasma-free in both infected (Figure.3.42.A, B and C) and uninfected plants (Figure.3.42.D).

No phytoplasma could be found in cells adjacent to the sieve elements including companion cells and phloem parenchyma in both infected (Figure.3.43.A, B and C) and uninfected plants (Figure.3.43.D). However, different organelles are present especially mitochondria. Mitochondria have a similar size to phytoplasma but could be clearly distinguished due to the presence of christae of the inner mitochondrial membrane ((Figure.3.42 and 43. A-D), arrows). Ultrathin sections show that bundle sheath and mesophyll tissues are also phytoplasma-free; (Figure.3.44).

In all cases, further electron microscopic studies should be carried out using immune-labelling on both light and electron microscopic level for phytoplasma specific proteins to exclude the presence of phytoplasma in companion cells. Due to the cytoplasma density a clear answer of the pathogen presence is not possible as in (Figure.3.42.B).

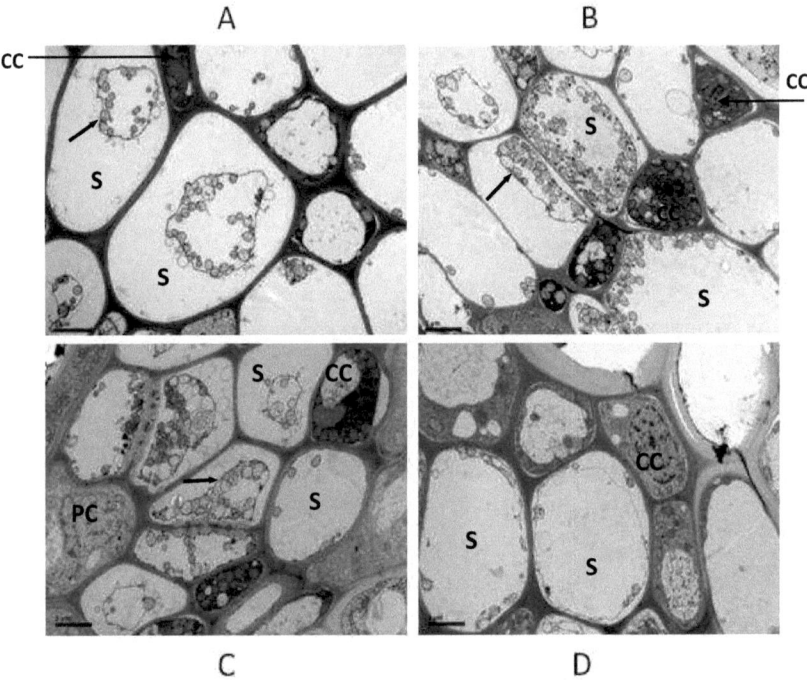

Figure.3.42. Comparison between phloem sieve elements of white, variegated and green leaves of white leaf phytoplasma-infected sugarcane. A: white leaf phenotype of sugarcane white leaf phytoplasma infection, where phytoplasmas are clearly visible in sieve elements (arrow). **B:** variegated leaf phenotype of sugarcane white leaf phytoplasma infection. Phytoplasmas are obvious present in sieve elements in increased numbers in comparison with white leaf (arrow). **C:** green leaf phenotype of sugarcane white leaf phytoplasma infection. Phytoplasmas are more abundant in some sieve elements than others (arrow). **D:** green leaf of uninfected sugarcane plant where it is used as control plant for comparison. It is obvious that no phytoplasma is present in sieve elements. Infected and uninfected leaves show the typical sieve anatomy. Vacuoles and cytoplasm are fused to a so called mictoplasm (Esau et al., 1965). Most of the organelles are absent, only typical round shaped sieve elements plastids are present. Sieve element plastids show the typical phenotype with crystal inclusions which are common for poaeceae (arrow Fig. 53 D). Sieve tube (S), companion cell (CC), parenchyma cell (PC). Bar= 2 µm.

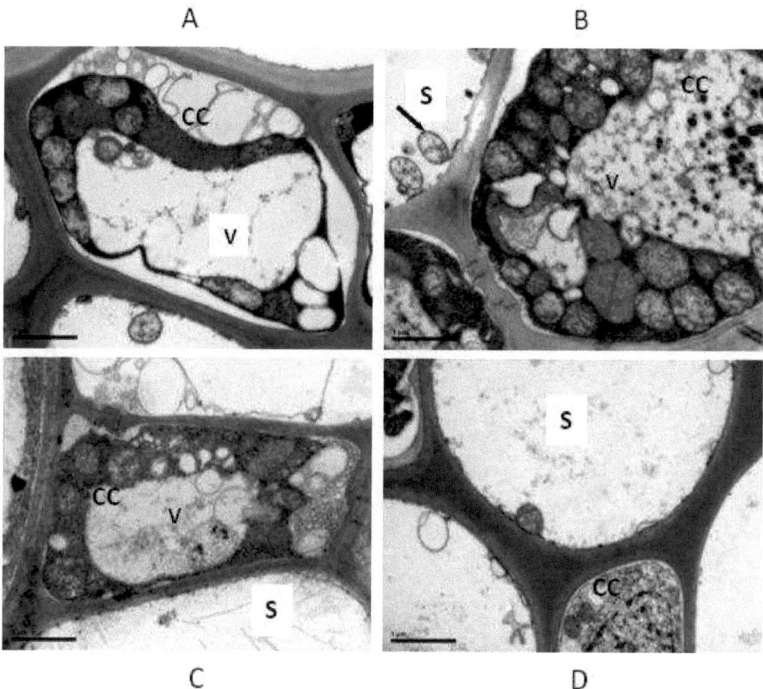

Figure.3.43. Comparison between phloem companion cells of phytoplasma infected and uninfected sugarcane. A: white leaf phenotype of phytoplasma-infected sugarcane where surrounding companion cell of sieve elements contains different organelles including vacuoles and mitochondria but they don't show any phytoplasma. B: variegated leaf phenotype of phytoplasma-infected sugarcane show typical companion cells connected to each other by plasmodesmata. Despite of these companion cells surround sieve element which contain phytoplasmas but these cells are phytoplasma-free. C: green leaf phenotype of phytoplasma infection where companion cell is also lack phytoplasma. D: green leaf of uninfected sugarcane which used as control and show partially one companion cell surrounds phytoplasma-free sieve element. Mitochondria have a similar size to phytoplasma but could be clearly distinguished due to the presence of Crystal of the inner mitochondrial membrane. Sieve tube (S), Companion cell (CC), vacuoles (V), plastid (P). Bar= 1 μm.

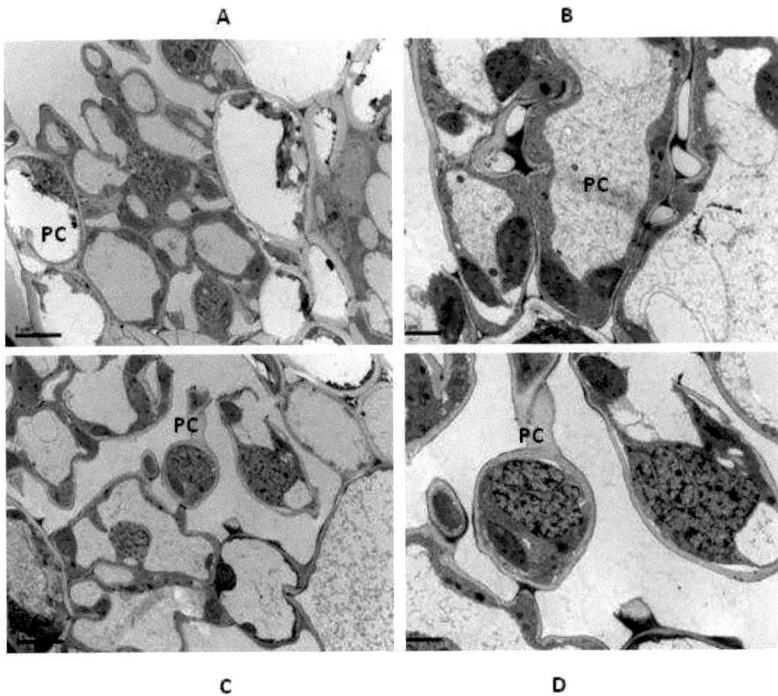

Figure.3.44. Ultrathin sections of bundle sheath and mesophyll tissues of phytoplasma-infected sugarcane.
A and B: variegated leaf phenotype of phytoplasma-infected sugarcane show that phytoplasmas are absent in the bundle sheath tissues. **C and D:** mesophyll tissues of variegated leaf phenotype of phytoplasma-infected sugarcane are also phytoplasma-free. Parenchyma cells (PC), (A), (C) bar = 5 µm. (B), (D) bar = 2 µm.

3.13.3. Phytoplasma size and shape

Size of the phytoplasma bodies varied from 200 nm to 800 nm (0.2 µm to 0.8 µm) in diameter. Our transmission electron microscopic studies of white leaf phytoplasma-infected sugarcane leaf showed spherical bodies which were bounded by a poorly defined membrane; (Figure.3.45). Sieve tubes filled with numerous phytoplasmas were seen particularly in variegated leaves of diseased sugarcane; (Figure.3.46).

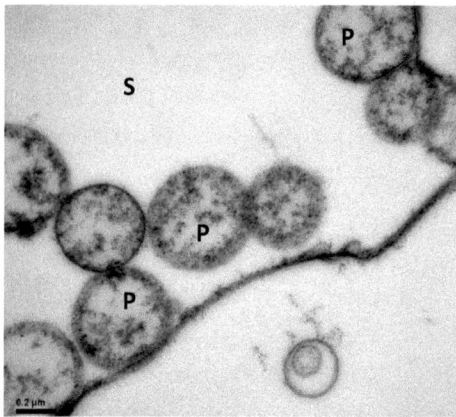

Figure.3.45. Transmission electron micrograph of typical membrane-bound phytoplasma bodies which present in sieve tube and contain resembling DNA in sieve tube of white leaf phytoplasma-infected sugarcane leaf. Sieve tube (S), phytoplasma (P). Bar = 0.2 µm.

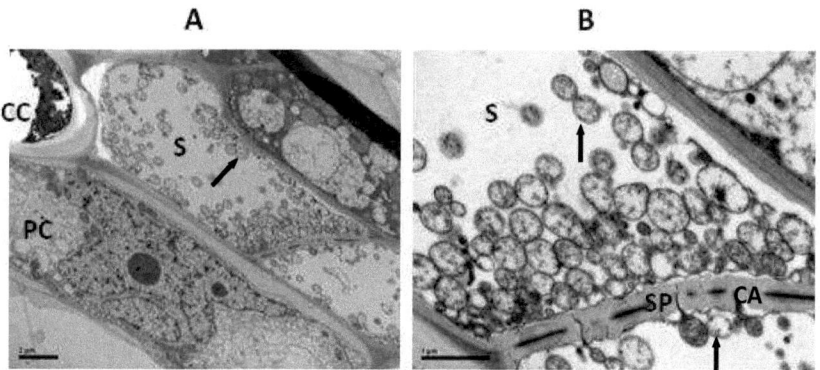

Figure.3.46. Ultrathin sections of phloem tissue of phytoplasma-infected sugarcane. A: Ultrathin section in variegated leaf phenotype shows that phytoplasmas (arrows) fill a phloem sieve element with a large number which is approximately more than 100 phytoplasma cells in one sieve tube. Phytoplasmas block the downward translocation photosynthates and passing through a sieve-plate pore lined with callose. B: higher magnification of the same last ultrathin section. Sieve cell (S), companion cell (CC), parenchyma cell (PC), sieve plate (sp), plastid (P), callose (CA). (A): bar = 2µm. (B): bar = 1µm.

Results

3.13.4. Ultrastructural changes of the phytoplasma infection on leave anatomy

Several ultrastructural changes were observed on ultrathin sections of the vascular tissues of affected sugarcane plants under transmission electron microscope (TEM). Paranchymatic cells of bundle sheath and mesophyll tissue of affected leaves showed some alterations comparing to uninfected leaves. In these cells accumulations of starch granules and plastoglobuli were observed in white leaf phytoplasma-infected sugarcane comparing to control uninfected one; (Figure.3.47). Our electron microscopic studies are in agreement with literature where phytoplasma infections led to a significant increase of starch in source leaves (Lepka et al. 1999). These data are consistent with ultrastructural observations reporting starch accumulation in chloroplasts associated with a severe disorganization of thylakoids and a reduction in chlorophyll content (Musetti, 2006).

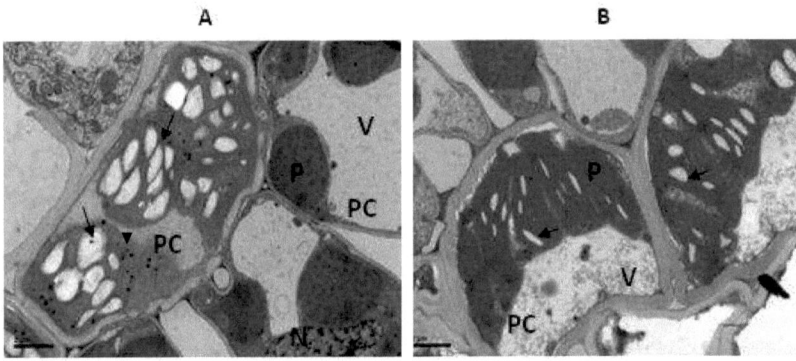

Figure.3.47. Ultrastructural comparison between paranchymatic bundle sheath cells of infected and uninfected sugarcane. A: variegated leaf phenotype of phytoplasma-infected sugarcane, where ultrastructural observations indicate accumulation of starch granules (arrows) and plastoglobuli (head arrow) in chloroplasts of bundle sheath cells of infected sugarcane. **B:** green leaf of uninfected sugarcane where it is used as control plant for comparison. It is clearly that accumulation of starch granules (arrows) in chloroplasts of bundle sheath cells is less than infected sugarcane. In addition, plastoglobuli are not accumulated in uninfected sugarcane. Parenchyma cell (PC), plastid (P), nucleus (N), vacuoles (V). Bar = 2μm.

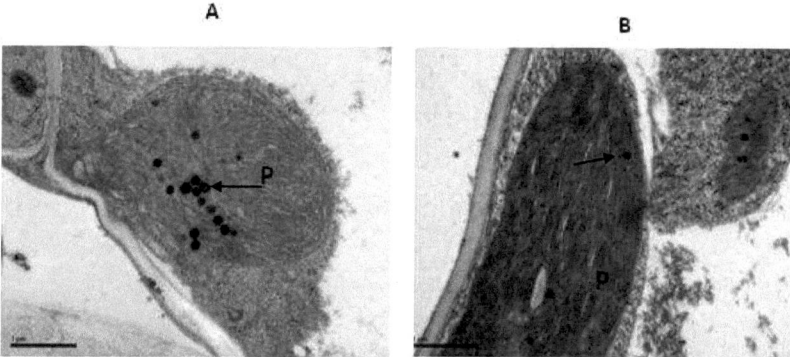

Figure.3.48. Comparison between chloroplasts structure of mesophyll cells in infected and uninfected sugarcane. A: variegated leaf of phytoplasma-infected sugarcane where an increase of plastoglobuli number and size in disorganized chloroplasts was found in mesophyll paranchymatic cells. **B:** green leaf of uninfected sugarcane which used as control, where mesophyll paranchymatic cells contain normal chloroplast with lower formation of plastoglobuli in comparison with infected sugarcane. Plastid (P), plastoglobuli (arrows). Bar = 1μm.

4. Discussion

4.1. Establishment of the test for phytoplasma

Yellow leaf syndrome (YLS) of sugarcane has been associated with several biotic and abiotic causes during the past four decades. Lute viruses and phytoplasmas are two types of plant pathogens that are typified as causing symptoms of yellowing in their hosts (Jones, 2002). It is hardly surprising that in sugarcane the pathogen cannot be distinguished by symptoms alone (Cronje et al., 1998; Arocha et al., 1999).

In the present study we have employed molecular-based tools for detection and identification of the putative causal agent of yellow leaf syndrome and report for the first time the presence of phytoplasma in Hawaiian sugarcane cultivars. The exceptional sensitivity of PCR offers many advantages for detection of plant pathogens (Herson and French, 1993). Application of this technique for detection and investigation of phytoplasmas seems particularly appropriate due to the small size of these plant pathogens and inability to culture them in vitro.

4.1.1. Efficiency of PCR amplification

Phytoplasma diagnostics and phylogenetics have historically been based on the 16S rRNA gene and the 16-23S rRNA spacer region because of the availability of universal primers for this region (Hodgetts and Dickinson, 2010). Numerous PCR primer combinations have been designed for diagnostics and phylogenetics. However, diagnostics based on these primers can be problematic, with occasional false positives, particularly through amplification of any *bacillusspp.* that might be present in a plant sample (Harrison et al., 2002). However, based on our investigation we never found *bacillus spp.* in sugarcane plant samples which was confirmed by RFLP analysis and DNA sequencing of nested-PCR products.

Though PCR is a routine technique for phytoplasma detection, there still meet some difficulties, at least with some primers which in some cases can induce dimmers, bands of non specific sizes. In these cases, false positive results can be expected. Two types of control were therefore routinely applied in each PCR run to test possible generation of false positive amplicons. One was the implementation of water control to test the generation of primer dimmers. However, in our hands, nested-PCR with all primer combinations used didn't amplify products from water used as template. The other control experiment was with sugarcane DNA from sugarcane plants which were phytoplasma-free, namely plants from

Egypt (Gt549, Ph8013) and plants which had undergone hot water treatment. These preparations did not give a phytoplasma-specific amplicon which indicated that the sugarcane DNA does not contain nucleotide sequences which bind to the phytoplasma-specific primers. In contrast, in many cases the same phytoplasma-infected sugarcane samples amplified with some primer pairs never reacted with other primer pairs, despite the fact that the used primers were universal. In the case of no visible products were obtained from phytoplasma positive samples, a higher dilution of DNA is used to dilute the plant inhibitors which may exist. However, these phytoplasma positive samples have shown false negative results after dilution too. It seems that in the case of phytoplasma positive samples, the primers preferentially amplified phytoplasma sequence of expected size. For example, (P1/P7, R16F2n/R16R2) and (R16mF2/R16mR1, R16F2n/R16R2) have been widely used for the detection of phytoplasma and are probably the most thoroughly investigated. They detect all strains of phytoplasmas whereas the DNA of non-infected plants does not react. Many phytoplasma positive samples were false negatives with these assays. Therefore, each sugarcane sample was investigated for phytoplasma by different nested-PCR assays (I), (II), (III)and(IV)with different primer pair combinations. Our tests have shown significant differences in the results of the PCR assays due of some weak or no amplification using particular primer combinations. According to our tests the primer pairs used for nested-PCR assay III, (MLO-X/MLO-Y) and (P1/P2), which amplified 16S-23S rRNA spacer region, was demonstrated to be the most reliable one to detect the phytoplasma in sugarcane plants due to the high efficiency of PCR amplification, high annealing temperature and low or no non-specific bands; in addition to the high sensitivity of these primer pairs where yielded standard products visualized in bands of a strong intensity. Our analysis demonstrated difficulties with the detection ability of phytoplasma in sugarcane plants. In order to explain the different result patterns obtained with particular primer combinations, a subliminal amount of template DNA, a presence of PCR-inhibiting substances in DNA preparations and sequential variability of primer target sites can be taken into account (Skrzeczkowski et al., 2001; Heinrich et al., 2001). For example, use of the 16S-23S rRNA spacer region in our investigation was more reliable than 16S rRNA gene region. It was more powerful than the 16S rRNA gene because it yielded standard products visualized in bands of a strong intensity as mentioned above.

As a consequence, in the case of critical samples, different primer pair combinations and also sequencing should be used for elucidation of phytoplasma presence (Franova, 2011).

Discussion

4.1.2. Carry-over contamination problems

The ability of the PCR to amplify minute amounts of template has the disadvantage that small quantities of contaminating DNA may be a problem for some applications like pathogens detection. In general, the titre of phytoplasma in sugarcane plants is very low and the standard method is nested PCR, which enhances the sensitivity of the test by two successive rounds of amplification. It is a very sensitive method and the risk of false positive signals is high. Two types of water control were therefore routinely applied to test for possible generation of false positive amplicons. One was the implementation of water control, where water was included in the first PCR-round instead of DNA from sugarcane leaves and then the hypothetical amplicon was transferred to the second PCR-round. In parallel the second PCR-round was also performed with water instead of the first-round amplicon and several water controls were used in each PCR round. On the other hand, it is important to have a designated clean area for setting up PCR reactions from which other DNA samples, especially PCR products, are excluded.

4.1.3. Q-PCR (real-time PCR)

Performance characteristics of used qPCR assay were determined by amplifying three separately prepared sets of dilution series of three standard samples in water. All systems gave good values as far as R^2, Ct, efficiency, limit of detection and sensitivity of amplicons which showed a broad dynamic range (five log orders of magnitude).

These parameters were also evaluated for the artificial samples that imitate infected sugarcane samples. The calibration curve of these artificial samples was very important to check sensitivity of real- time PCR assay and to explain the false negative results of qPCR for most our sugarcane samples. The irregular signals at the sixth dilution of artificial samples look similar to the signals of experimental sugarcane samples (Figure 4.1). It is most likely that at sixth concentration level (10^{-5}), using a 10-fold dilution series of the phytoplasma-infected sugarcane (phytoplasmal DNA) mixed with sugarcane DNA to imitate real sugarcane samples, the titer of phytoplasmal DNA is very low. It could be that sugarcane samples, which contain low titer of phytoplasmas cannot be detected sensitively by this direct qPCR assay due to the influence of host-material and that may be true for the sugarcane plants which show yellow leaf syndrome and contain low titer of phytoplasma.

Discussion

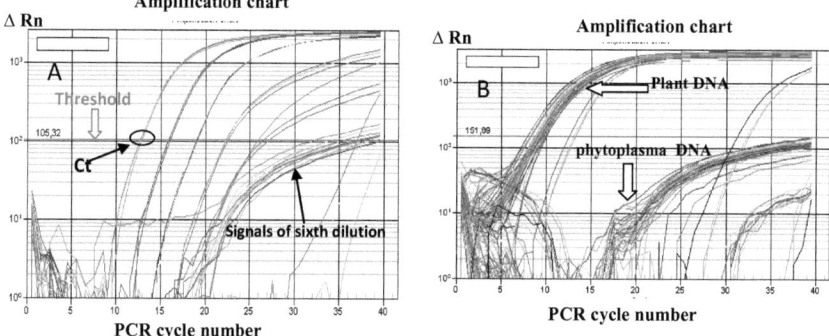

Figure.4.1. A: Log- view of standard curve chart. Standard curve determined at six concentration levels (ranging from 10^0 to 10^{-5}) using 10-fold dilution series of the phytoplasma-infected sugarcane (phytoplasmal DNA) mixed with sugarcane DNA to imitate real sugarcane samples. **Threshold;** is an arbitrary level of fluorescence chosen on the basis of the baseline variability. **Ct;** is defined as the fractional PCR cycle number at which the reporter fluorescence is greater than the threshold. ΔRun; is an increment of fluorescent signal at each time point. The ΔRun values are plotted versus the cycle number. **B: Log- view of amplification chart of qPCR results.**

On the other hand, it is essential that the nucleic acid is sufficiently pure for qPCR analysis. Template contamination (i.e., protein, carbohydrates or organic solvents) can have a huge impact on assay reliability and reproducibility. We used high pure PCR template preparation kit. Then the template DNA quality was determined by Nanophotometer. Since, diagnosis of pathogen in the plants including sugarcane is often hampered by the presence of PCR inhibitors such as polyphenolics, polysaccharides and other molecules that may produce false-negative results even sometimes from heavily infected samples (Weintraub and Jones, 2010). To prove that the absence of a signal is not due to such causes, protocols for control amplification and detection of the host plant DNA have been developed such as 18S rRNA gene (Christensen et al., 2004).

According to our plant 18S rRNA gene analysis, however, sugarcane samples were sufficiently pure for qPCR analysis. It appears that PCR inhibitors can hamper diagnosis of phytoplasma only when phytoplasmas exist in very low titer as most of our sugarcane samples.

Discussion

Our analysis of field-collected sugarcane samples showed significantly higher diagnostic sensitivity of conventional nested-PCR in comparison to direct qPCR assay by revealing most false negative PCR results. Most samples that were detected as phytoplasma-negative by qPCR were shown to be positive by nested-PCR. It is most likely that this qPCR assay can detect the phytoplasma only in heavily infected sugarcane samples which are showing strong symptoms like sugarcane white leaf and sugarcane grassy shoot. On the other hand, the primer pair and probe which were used in this qPCR assay can theoretically anneal to nucleotide sequences of the phytoplasma strains which infect our experimental sugarcane samples (Figure.4.2). It is more likely that the false negative results are attributed to the level of phytoplasma infection than to the phytoplasma strain differences.

```
TAAAGACCTTTTTCGGAAGGTATGCTTAAAGAGGGGCTTGCGGCACATTAGTTAGTTGGTAGGGTAAAGGCCT
ACCAAGACTATGATGTGTAGCTGGACTGAGAGGTTGAACAGCCACATTGGGACTGAGACACGGCCCAAACTCC
TACGGGAGGCAGCAGTAGGGAATTTTCGGCAATGGAGGAAACTCTGACCGAGCAACGCCGCGTGAACGATGA
AGTATTTCGGTATGTAAAGTTCTTTTATTGAAGAAGAAAAAATAGTGGAAAAACTATCTTGACGCTATTCAATG
AATAAGCCCCGGCAAACTATGTGCCAGCAGCCGCGGTAATACATAGGGGGCAAGCGTTATCCGGAATTATTGG
GCGTAAAGGGTGCGTAGGCGGTTTAATAAGTCTATAGTTTAATTTCAGTGCTTAACACTGTTCTGCTATAGAAA
CTATTAAACTGGAGTGAGATAGAGGCAAGTGGAATTCCATGTGTAGCGGTAAAATGCGTAAATATATGGAGGA
ACACCAGAGGCGTAGGCGGCTTGCTGGGTCTTTACTGACGCTGAGGCACGAAAGCGTGGGGAGCAAACAGGA
TTAGATACCCTGGTAGTCCACGCCGTAAACGATGAGTACTAAGTGTCGGGGTTACTCGGTACTGAAGTTAACA
CATTAAGTACTCCGCCTGAGTAGTACGTACGCAAGTATGAAACTTAAAGGAATTGACGGGACTCCGCACAAGC
GGTGGATCATGTTGTTTAATTCGAAGATACACGAAAAACCTTACCAGGTCTTGACATACTCTGCAAAGCTATAG
CAATATAGTGGAGGTTATCAGGGATACAGGTGGTGGCATGGTTGTCGTCAGCTCGTGTCGTGAGATGTTAGGTT
AAGTCCTAAAACGAGCGCAACCCTTGTCACTAGTTGCCAGCATGTTATGATGGGCACTTTAGTGAGACTGCCA
ATGAAAAATTGGAGGAAGGTGAGGATCACGTCAAATCATCATGCCCCTTATGATCTGGGCTACAAACGTGATA
CAATGGCTGTTACAAAGAGTAGCTGAAACGCAAGTTAATAGCCAATCTCATAAAAACAGTCTCAGTTCGGATT
GAAGTCTGCAACTCGACTTCATGAAGTTGGAATCGCTAGTAATCGCGAAT
```

Figure.4.2.Binding sites of qPCR primers and probe on the nucleotide of partial 16S rRNA gene sequence of SCGS phytoplasma infects Egyptian sugarcane cultivar (G8447). This sample was detected as phytoplasma-negative by qPCR but shown to be positive by nested-PCR. Binding site of forward primer is marked with yellow colour; binding site of reverse primer is marked with green colour, whereas binding site of probe is marked with blue colour.

In all cases, other qPCR assays may be carried out using primers and probes designed based on the other regions like 23S rRNA gene or designed based on the different gene (non-ribosomal) like *tuf* gene in order to test if other direct qPCR assays can detect the phytoplasma in sugarcane plants better and therefore, more reliable. As mentioned above our results based on nested-PCR assays demonstrated also difficulties with the detection of

Discussion

phytoplasma in sugarcane plants where significant differences in results of nested-PCR assays occurred using particular primer combinations.

4.2. Phytoplasma detection and identification by RFLP analysis

Restriction fragment length polymorphism (RFLP) analysis of PCR-amplified rRNA gene is the preferred method to identify and differentiate phytoplasmas within primary phylogenetic groups that were established by 16S rRNA gene sequence analysis. This straightforward method was introduced into phytoplasmology by Lee et al., (1998); Schneider et al., (1993).
The comprehensive classification scheme, combined with illustrative RFLP patterns characteristic of each distinct group and subgroup, continues to provide a simple, reliable and practical means to identify unknown phytoplasmas without the need to sequence the 16S rRNA gene.

4.2.1. Phytoplasma types in Hawaiian and Cuban sugarcane

DNA from sugarcane cultivars, which had been obtained from the Hawaiian breeding station (2003) and a Cuban station (2005), gave an amplicon which was similar in size and in restriction fragment pattern to positive controls from phytoplasma-infected periwinkle. Two types of phytoplasma were identified in these cultivars; two Cuban cultivars contained X-disease phytoplasma"*Ca.* Phytoplasma pruni", whereas the 6 Hawaiian cultivars and one Cuban cultivar contained the Aster yellows phytoplasma"*Ca.* Phytoplasma asteris" (Figure.3.4and Table.3.2). In addition, faint bands were also identified in some Hawaiian cultivars and suggested a mixed population of phytoplasma in sugarcane white leaf disease (Figure.3.4). The most likely explanation for the banding pattern of PCR amplified DNA from sugarcane white leaf (SCWL) is dual infections by different phytoplasma groups.
 Multiple phytoplasma infection based on an analysis of the banding pattern of restriction enzyme-digested PCR products has been reported in sesame plants (Nakashima et al., 1996) and grapevines (Bianco et al., 1993).

Further surveys were carried out in order to show how widespread phytoplasma may be in plants from previous sugarcane fields and whether there are fluctuations of the phytoplasma infection in the breeding station due to pesticide treatments. Therefore, sugarcane samples were collected from Hawaiian plantations in 2009 and 2011 and from breeding station in 2010 and 2011 and conserved as dried leaves samples. Our investigations revealed that there are

Discussion

two types of phytoplasmas: Sugarcane white leaf phytoplasma as predominant group and aster yellows group.

It is not immediately obvious why this diversity of phytoplasmas exists in Hawaiian sugarcane, despite the case of diversity of phytoplasmas in northern Australian sugarcane was previously reported (Tran-Nguyen et al., 2000). On the other hand, closely related phytoplasmas have been reported to induce different phenotypic symptoms and different types of phytoplasmas to induce similar symptoms (Davis et al., 1998; Firrao et al., 2005). It could be the later is the case of Hawaiian phytoplasma infected-sugarcane.

Aster yellows (16SrI) phytoplasmas are an example of one of the most diverse phytoplasmas known to date. Aster yellow phytoplasma: originally described in aster (Asteraceae family), this phytoplasma infects herbaceous plants in over forty families, including vegetables, ornamentals and weeds. Aster yellows phytoplasma type in sugarcane had already been reported in Cuban sugarcane as well as X-disease phytoplasma (Arocha et al., 1999).

The only phytoplasma infection found in Hawaii so far was the infection of water cress by an Aster yellows type phytoplasma (Borth et al., 2002) and of *Dodonea* by a Western-X-disease phytoplasma (Borth et al., 1995). No phytoplasma in Hawaiian sugarcane was reported so far. The close relationship between the phytoplasma strain found by us in Hawaiian sugarcane and the one reported in water cress from Hawaii poses the question, whether there had been a transfection from water cress to sugarcane, although according to literature the specificity of phytoplasmas and of their vectors excludes transfection from dicots to grasses. The specificity even differentiates between different grass species including sugarcane (Tran Nguyen et al., 2000; Wongkaew et al., 1997). The Hawaiian sugarcane breeding station (from where the plants in the Bayreuth green house had been obtained) never had water cress fields in their proximity. Furthermore many phytoplasma strains from very different hosts show close similarity to the Hawaiian SCYLP, so that the similarity to water cress phytoplasma may not imply that the strains are transmissible between these different hosts.

We wanted to do more surveys in order to check whether sugarcane plants, which grow close to water cress fields, contain the water cress-typical phytoplasma, but phytoplasma-infected water cress plants were not available; due to the reason that all phytoplasma-infected water cress plants had already been eliminated. For example, the Sumida water cress field had phytoplasma-infected water cress 10 years ago and these plants had been eliminated. Furthermore, currently there are no sugarcane plants nearby this field. Some of perennial

Discussion

grasses (some of them are wild relatives of sugarcane such as *Miscanthus Sp.*)still exist nearby this field which may be a reservoir of phytoplasma. According to our investigation, these plants are not infected with phytoplasma.

It is also not immediately obvious how sugarcane white leaf (SCWL) phytoplasma exists in Hawaiian sugarcane, but it suggests that there is active insect transmission occurring either within the Hawaiian Islands or from outside the Hawaiian Islands. Since, phytoplasma diseases of sugarcane including sugarcane white leaf (SCWL)and sugarcane grassy shoot (SCGS) have been reported to cause substantial losses in the sugarcane crop all over Asia (Chona et al., 1960; Chen, 1974; Rishi and Chen, 1989) and the chance of their transmission to other geographical regions seems to be high, given the large phytoplasma reservoir already revealed, the propensity of new phytoplasma strains to evolve (Lee et al., 2000) and the ability of leafhoppers, the most common insect vectors of phytoplasma, to migrate long distances (Wongkaew, 1999; Viswanathan, 2000) and switch to new host plants (Purcell, 1985).

Another possibility is that sugarcane white leaf phytoplasma already existed in mother plant cuttings which were obtained from Asia before they were propagated for use on Hawaiian Islands. Since, Hawaiian sugarcane phytoplasma isolate (JN223448) from cultivar H78-7750 ,which was obtained from Hawaiian breeding station, clustered to (SCWL) phytoplasma closely together with (SCWL) phytoplasma from Taiwan (AY139874) with a bootstrap value of 86.6and shared 98% sequence identity (Table.3.29 and Figure.3.35.b). This is the first report indicating an association of sugarcane white leaf (SCWL) phytoplasma strain with sugarcane plants showing yellow leaf syndrome symptoms.

4.2.2. Phytoplasma types in Thai sugarcane

Sugarcane (*Saccharum* sp. and hybrids) is affected by two lethal phytoplasmal diseases, i.e., sugarcane grassy shoot (SCGS) and sugarcane white leaf (SCWL) (Rao et al., 2005). SCGS disease has been reported to occur in India, Bangladesh, Malaysia, Nepal and Pakistan whereas SCWL is predominant in Taiwan, Sri Lanka and Thailand (Rao et al., 2005). They are caused by SCGS and SCWL phytoplasmas, respectively. These two phytoplasmas belong to rice yellow dwarf group"*Ca.* Phytoplasma oryzae"also named sugarcane white leaf (SCWL) group (Jung et al., 2003; Marcone et al., 2004).

Discussion

Our analysis showed that Thai sugarcane (unknown cultivar) from province of Khon kaen contains phytoplasma isolate(HQ917068) clustered to the rice yellow dwarf group, closely together with sugarcane white leaf phytoplasma from Myanmar (AB646271) with a bootstrap value of 64.2and shared 100% sequence identity (Table.3.29 and Figure.3.35.b).

Our results showed that other Thai sugarcane plant samples from province of Suphan Buri contain also SCWL phytoplasma. However, these plants showed yellow leaf syndrome symptoms but not white leaf symptoms (Figure.1.7) and that is also true for many Hawaiian sugarcane plants where yellow leaf syndrome symptoms are associated with SCWL phytoplasma type as mentioned before.

Based on these results, SCWL phytoplasma may be associated with sugarcane plants showing yellow leaf syndrome symptoms. In fact, it is well known that SCWL phytoplasma is responsible for sugarcane white leaf disease symptoms but association of SCWL phytoplasma type with sugarcane yellow leaf syndrome symptoms was not previously documented. It could be that these phytoplasmas cause yellow leaf syndrome symptoms when they exist in low titer as in the case of Hawaiian sugarcane and Thai sugarcane from province of Suphan Buri. On the other hand, SCWL phytoplasma belongs to rice yellow dwarf group and this type of phytoplasma causes the yellowing of the infected rice.

4.2.3. Phytoplasma types in Egyptian and Syrian sugarcane

According to RFLP profiles of HinfI restriction enzyme, Egyptian sugarcane cultivar G8447 contains phytoplasma, which belongs to the rice yellow dwarf group and strain of sugarcane grassy shoot.

Nucleotide sequence analysis of 16S rRNA genes revealed that sugarcane grassy shoot (SCGS) phytoplasma is very closely related to the sugarcane white leaf (SCWL) agent and sharing a sequence similarity more than 98%. However, strains of SCGS phytoplasma for which the full-length 16S rRNA gene sequences are available from GenBanks, lack the HinfI restriction site, which is present in the 16S rRNA gene sequences of strains SCWL agent (Govindet al., 2007). Therefore, the SCGS phytoplasma can be distinguished from the SCWL agent using RFLP analysis with HinfI restriction endonuclease (Hanboonsong et al., 2002; Marcone et al., 2004). In contrast, at the 16S–23S rRNA gene spacer sequence level, SCWL and SCGS phytoplasma isolates are identical or nearly identical. Therefore, 16S–23S rRNA gene spacer sequence is a less significant taxonomic tool than 16S rRNA gene sequence.

Discussion

To our knowledge, this is the first report about the presence of SCGS phytoplasma in Egyptian sugarcane cultivar. The geographical distribution of grassy shoot (GS) includes countries such as Bangladesh, India, Iran, Malaysia, Myanmar, Nepal, Pakistan, Sri Lanka, Sudan and Thailand (Viswanathan, 2000).

Association of phytoplasma with streak yellows on date palm in Egypt was documented for the first time in 2005 (Ammar et al., 2005). In addition, occurrence of phytoplasma diseases in Egypt was reported in periwinkle and was identified as members of aster yellow phytoplasma group (Omar et al., 2008). No phytoplasma in Egyptian sugarcane was documented so far.

According to DNA sequence analyses of nested-PCR products of Syrian sugarcane cultivar using P1/P2 primer pair, this plant contains phytoplasma, which falls into the rice yellow dwarf (SCWL) group. However, rely on RFLP profiles with restriction endonucleas (HinfI) this Syrian sugarcane cultivar contains non-identified phytoplasma strain (Figure.3.9 and 3.10).Therefore, other DNA sequence analyses using other primer pairs are required in order to identify this phytoplasma isolate. This is also true for some Thai sugarcane cultivars from province of Bang Phra which contains the same non-identified phytoplasma strain too.

Apple proliferation phytoplasma, which infects apple trees, was reported in south Syria. Also two types of phytoplasmas were identified in mixed infection in grapevine in Syria: one related to stolbur (16SrXII) and the other tentatively related to clover proliferation group (16SrVI) (Contaldo et al., 2011) but no phytoplasma in Syrian sugarcane was reported before.

As a consequence, our results are in agreement with those described in Australia, South Africa, Cuba and Mauritius (Vega et al., 1997; Cronje et al., 1998; Arocha et al.1999; Aljanabi et al., 2001) where phytoplasmas have been associated with YLS of sugarcane plants. Moreover, sugarcane yellow leaf syndrome (SCYLS) disease, which has been reported from several African countries, Cuba and Australia was associated with distinctly different phytoplasmas which are not specific pathogens. They include members of the X-disease, faba bean phyllody, aster yellows and SCWL groups which are known to infect a wide range of wild and cultivated plants and have low insect vector specificity (Marcone, 2002).

Discussion

4.3. Identification of the phytoplasma strains in sugarcane by phylogentic analysis

Due to the inability to cultivate phytoplasmas in cell-free media, molecular analyses of conserved gene sequences have become rational means for phytoplasma taxonomy and classification. Use of DNA sequences to build up phylogenetic trees is widespread and recognized as a valid approach for identifying taxonomic relationships between organisms (Hodgetts and Dickinson, 2010).

Following decisions for phytoplasma taxonomy taken by the Phytoplasma Working Team during the 13th International Organization of Mycoplasmology held in Fukuoka, Japan (14 to 19 July 2000) in general, a strain can be described as a new "*Candidatus* Phytoplasma species" if its 16S rRNA gene sequence has less than 97.5% identity to any previously described "*Candidatus* Phytoplasma species."

A BLAST search for the 16S rRNA gene sequences reported in this study showed that they shared 99 to 100% sequence identity with those of other phytoplasmas in the aster yellows, X-disease and rice yellow dwarf groups. This confirmed that the detected phytoplasmas belong to these groups of '*Candidatus phytoplasma*'. For example, Egyptian sugarcane cultivar G8447 contains phytoplasma strain (JN223446)clustered to the rice yellow dwarf group, closely together with sorghum grassy shoot phytoplasma from Australia (AF509324) with a bootstrap value of 81.1 and shared 99% sequence identity (Table.3.29 and Figure.3.35.a). Furthermore, it was previously reported that the more distantly related to SCGS agent, is the sorghum grassy shoot (SGS) (Rao et al., 2007). Therefore, the Egyptian sugarcane cultivar G8447 contains phytoplasma strain (JN223446)belongs to the rice yellow dwarf group '*Candidatus phytoplasma oryzae*' and this strain cannot be described as a new "*Candidatus* Phytoplasma species" due of its 16S rRNA gene sequence has more than 97.5% identity to any previously described "*Candidatus* Phytoplasma species.". That is also true for phytoplasma isolate (HM804282)from Cuban sugarcane cultivar Ja605 clustered together with other strains of X-disease group, among them already reported sugarcane yellows phytoplasma strain found in South Africa (AF056095) with a bootstrap value of 48.7and shared 99% sequence identity (Table.3.29 and Figure.3.35.a). Therefore, the Cuban sugarcane cultivar Ja605 contains phytoplasma strain (HM804282)belongs to the X-disease group '*Candidatus phytoplasma* pruni' and can't be described as a new "*Candidatus* Phytoplasma species".

Other Cuban sugarcane cultivar C10-5173 was infected with phytoplasma strain (HQ116553) clustered to the aster yellows group, closely together with sugarcane yellows phytoplasma

Discussion

from Brazil (EU423900) and maize bushy stunt phytoplasma from Colombia (HQ530152) with a bootstrap value of 49.6 and shared 99% sequence identity (Table.3.29 and Figure.3.35.a). Therefore, the Cuban sugarcane cultivar C10-5173 contains phytoplasma strain (HQ116553) belongs to the Aster yellows group '*Candidatus phytoplasma* asteris' and this strain can't be also described as a new "*Candidatus* Phytoplasma species".

4.4. Hot water treatment in order to get phytoplasma free plant

When a pathogen is excluded from the propagating material of a host, it is often possible to grow the host free of that pathogen for the rest of its life (Aslam, 2001).

Hot water treatment has been proposed to cure dormant woody plant material from phytoplasmas. While tissue culture techniques are routinely used for virus eradication; few reports have been published on their potentiality in phytoplasma elimination (Dai et al., 1997; Parmessur et al., 2002; Chalak et al., 2005). The effectiveness of the method is based on the fact that dormant plant organs can withstand higher temperatures than those their respective pathogens can survive for a given time (Agrios, 2004).

The first aim of our hot water treatment was to get negative control (phytoplasma-free plant) Therefore, we used an Australian recipe (Arocha, 2005b), as long duration treatment (48 h in cold water (10°C), followed by 3h in hot water (50°C)).The cuttings (approx. 30 mm average diameter) were kept in cold water before the hot water treatment was applied.

It seems that, immersion at 50°C for 30 min was not effective to eliminate the phytoplasma totally from the cuttings but it could be effective in smaller diameters. Furthermore, it could be that also depends on the titer of phytoplasmas in the plant material.

Our tests showed that the appropriate hot water treatment, which recommended for phytoplasma elimination, is immersion at 50°C for at least 60 min. Furthermore, our tests showed that the plant material (cuttings) should be thermally prepared to the treatment by storage for 48 hours at 10°C in order to prevent a poor vegetative development especially for long duration treatment at 50°C for 3 hours. Due it could be that this treatment could lead to high mortality rates.

The hot water treatment which is practiced by the Hawaiian plantations for their seed cane fields (3 h at 50°C) is sufficient to eliminate phytoplasma, whereas the duration of the hot water treatment for seed pieces which are planted in the fields (20 min at 52°C) may be at the margin of successful bacteria elimination.

Discussion

Therefore field plants were tested for phytoplasma with emphasis on the comparison of green plants with YLS-symptomatic plants, standing side-by-side. Our results showed that these field plants were in the case of a mostly phytoplasma-infected and it seems that there is no clear association between phytoplasma and symptoms due to some green sugarcane plants were also positive for phytoplasma. Asymptomatic sugarcane was frequently phytoplasma-positive; this has also been reported by other workers (Cronje et al., 1998a).

One explanation for the poor correlation between phytoplasma and symptoms is that some phytoplasmas can exist in plants without ever causing disease or having only a minor impact, as is the case for ash yellows in velvet ash (Sinclair et al., 1994) and phytoplasmas in alders (Lederer and Seemuller, 1991), apricots (Kirkpatrick et al., 1990) and almonds (Uyemoto et al., 1992). Such associations suggest that the host plant is either tolerant or resistant to phytoplasma infection.

4.5. Transmission test with sugarcane aphid

Our tests showed that the sugarcane aphids (*Melanaphis sacchari*) are able to acquire the phytoplasmas because DNA extracted from these insects produced an expected size of nested-PCR product. In addition, DNA sequencing of these PCR products confirmed that. Our tests showed that these sugarcane aphids are able to acquire the phytoplasmas but they are unable to transmit the phytoplasmas into the sugarcane plants because all target plants (phytoplasma-free plants) were negative for phytoplasma infection after three months post inoculation.

In fact, many aphids, whiteflies and mealy bugs are phloem-feeders on plant species infected with phytoplasmas, but so far none of them has been found to be a vector of phytoplasmas. Recently, apple aphids were found to be positive in PCR assays for apple proliferation phytoplasmas and were suspected to be vectors, but the results of transmission experiments seem to exclude this possibility (Cainelli et al., 2007). A phloem-feeding habit is thus necessary but insufficient for phytoplasma transmission.

It was an expected result that sugarcane aphids are unable to transmit the phytoplasmas (SCYP) because thus far, there has been no report of phytoplasma or spiroplasma transmission by a phloem-feeding aphid. The reasons for lack of transmission by aphids are not known. Sites of mollicute attachment to insect tissues and other pathogen-insect interactions can be cited in a general sense to explain transmission specificities. But what molecular mechanisms, that are present in leafhoppers and presumably absent in aphids,

Discussion

account for the differences in mollicute transmission between these major insect groups? (Mishra, 2004). As a consequence, interestingly, aphids apparently do not serve as phytoplasma vectors.

4.6. Transmission electron microscopy for cytological location of phytoplasma

Phytoplasmas are transferred with saliva of infected insect vectors into the pierced sieve element, from which they spread systemically in the plant using the continuous sieve tube system due they are pleiomorfic and sufficiently small to pass freely through sieve pores.

Our transmission electron microscopic studies revealed the presence of sugarcane white leaf phytoplasma only in phloem sieve tubes of diseased sugarcane leaves but not in cells adjacent to the sieve elements including companion cells and phloem parenchyma, although in many cases the phytoplasmas have been reliably documented in companion cells and phloem parenchyma cells by electron microscopy, as well as in sieve elements (Siller et al., 1987).

Several ultrastructural changes were observed under transmission electron microscope (TEM). Parenchymatic cells of bundle sheath and mesophyll tissue of infected leaves showed some alterations compared to uninfected leaves. In these cells accumulations of starch granules and plastoglobuli were observed in white leaf phytoplasma-infected sugarcane compared to uninfected control (Figure.3.47). Our electron microscopic studies are in agreement with literature, where phytoplasma infections led to a significant increase of starch in source leaves (Lepka et al., 1999). These data are consistent with ultrastructural observations reporting starch accumulation in chloroplasts associated with a severe disorganization of thylakoids and a reduction in chlorophyll content (Musetti, 2006) due to the decrease of both Chl a and Chl b in leaves. "A decrease in photosynthetic pigments has been observed in maize plants infected with maize bushy stunt (Junqueira et al., 2004), apples infected with apple proliferation and grapevine infected with the bois noir phytoplasma. This is probably the result of enhanced chlorophyllase activity in infected leaves (Bertamini et al., 2002b) and it has been suggested that phytoplasmas have a role in the inhibition of chlorophyll biosynthesis in plant host leaves (Bertamini et al., 2002a)."

"The descent of photosynthesis is the result of phytoplasma infection on photosynthetic electron transport and enzymatic activities, due to the loss of several thylakoid membrane proteins and to the reduction of leaf soluble proteins. These changes are similar to those induced by leaf ageing, so an interference of phytoplasmas with plant hormones that regulate senescence processes in leaf tissues could be hypothesized. In all kinds of diseases in which there is destruction of leaf tissue like sugarcane white leaf phytoplasma, photosynthesis is reduced because the photosynthetic surface of the plant is lessened. Most viruses, mollicute diseases induce varying degrees of chlorosis and stunting. In the majority of such diseases, the photosynthesis of infected plants is reduced greatly. In diseases caused by phytoplasmas, bacteria exist and reproduce in the phloem sieve tubes, thereby interfering with the downward translocation of nutrients" (Musetti, 2006).

An increase plastoglobuli number and size in disorganized chloroplasts was found in mesophyll paranchymatic cells of variegated leaf of white leaf phytoplasma-infected sugarcane compared to green leaf of uninfected sugarcane which was used as control (Figure.3.48). It is known that characteristics of plastids in senescent cells include reduced size, rounded shape and larger plastoglobuli (Thomson and Platt-Aloia, 1987; Biswal and Biswal, 1988; Noode´n, 1988).

As a consequence, the phytoplasma diseases are complex and their progress is also highly variable and depends on many factors including the state of the host plants, the pathogen and its different biotypes, the tendency for mutation, the presence and dynamics of the vectors, the titer of the phytoplasma, the environmental conditions as well as the agronomical practices being used (Ciancio and Mukerji, 2008).

5. Summary

The Yellow leaf syndrome (YLS) had been first detected and described in Hawaiian sugarcane plantations. The polerovirus *Sugarcane yellow leaf virus* was identified as a causal agent of the syndrome; however there was no strict correlation between the degree of symptom expression and the virus titre. Therefore several surveys on breeding station sugarcane plants in Hawaiian Islands were done for *Sugarcane yellow leaf phytoplasma* (SCYLP), a bacterium which had been hypothesized to be also a causal agent of YLS.

Two types of phytoplasmas were found in Hawaiian sugarcane cultivars mainly sugarcane white leaf phytoplasma (SCWL) which is a member in rice yellow dwarf group, in addition to aster yellows group. This was also true for sugarcane plants from Hawaiian plantations, which routinely use hot water-treatment for the seed cane cuttings.

Sugarcane samples were obtained also from other countries including Cuba, Egypt, Syria and Thailand where sugarcane plants are also showing symptoms of yellowing or whiting. Aster yellows and X-disease phytoplasmas were found in Cuban cultivars whereas one sugarcane cultivar from Egypt contains grassy shoot phytoplasma that is a member in rice yellow dwarf group, but the other two Egyptian ones were phytoplasma-free. Syrian sugarcane was infected by phytoplasma that identified preliminary in rice yellow dwarf group. To our knowledge, this is the first report for the detection and identification of phytoplasma in sugarcane plants from Hawaii, Egypt and Syria. Our investigation on Thai sugarcane plants was in agreement with previous literature where sugarcane white leaf (SCWL) phytoplasma is associated with white leaf disease (Nakashima et al., 1994; Wongkaew et al., 1997).

Q-PCR (real-time PCR) offers the opportunity to detect the phytoplasma in a sensitive, specific and quick manner, but that is not true for sugarcane plants with a very low titer of phytoplasma. Therefore, nested-PCR is better than qPCR for low titer phytoplasma detection and that is true for sugarcane yellow leaf phytoplasma disease. A BLAST search for the 16S rRNA gene sequences reported in this study showed that they shared 99 to 100% sequence identity with those of other phytoplasmas in the Aster yellows, X-disease and Rice yellow dwarf groups. However, no one of these identified strains can be described as a new "*Candidatus* Phytoplasma species". On the other hand, Hawaiian sugarcane cultivar H78-7750as a representative of Hawaiian breeding station sugarcane contains phytoplasma clustered to strain sugarcane white leaf (SCWL) phytoplasma, closely together with sugarcane white leaf phytoplasma from Taiwan (AY139874). It is possible to explain the occurrence of (SCWL) phytoplasma in Hawaiian Islands, by insect vectors or by infected stem cuttings which were obtained from other countries. Thai sugarcane contains phytoplasma isolate closely together with sugarcane white leaf phytoplasma from Myanmar.

The transmission electron microscopic (TEM) studies revealed the presence of sugarcane white leaf phytoplasma only in phloem sieve tubes of diseased sugarcane leaves but not in adjacent cells to the sieve elements including companion cells and phloem parenchyma as well. According to ultrastructural observations under TEM, parenchymatic cells of bundle sheath and mesophyll tissue of affected leaves showed some alterations including accumulations of starch granules, increase plastoglobuli number and size in disorganized chloroplasts.

Insect vectors of phytoplasmas are phloem feeders. Thus far, none of aphid species has been found to be a vector of phytoplasmas. Our tests showed also that black sugarcane aphids (*Melanaphis Sacchari*) were unable to transmit the phytoplasmas from infected sugarcane into the phytoplasma-free one. Hot water treatment has been proposed to cure plant material from phytoplasmas. Our tests showed that the appropriate hot water treatment, which recommended for phytoplasma elimination, is immersion of the sugarcane stem cuttings at 50°C for 60 min.

6. Zusammenfassung

Das Yellow Leaf Syndrom (YLS) bei Zuckerrohr wurde zuerst in Plantagen Hawaiis entdeckt und von dort beschrieben. Das Polerovirus *Sugarcane Yellow Leaf Virus* konnte als verursachendes Agens des Syndroms identifiziert werden, jedoch gab es keinen strikten Zusammenhang zwischen der Intensität der Symptome und dem Virustiter. Deshalb wurden Analysen an Zuckerrohrpflanzen aus der hawaiianischen Zuchtstation durchgeführt, um die Pflanzen auf *Sugarcane yellow leaf phytoplasma* (SCYLP) zu testen, einem Bakterium, das ebenfalls als möglicher Auslöser von YLS vermutet wurde.

Zwei Typen von Phytoplasma wurden in den hawaiianischen Zuckerrohrkultivaren entdeckt, nämlich *Sugarcane White Leaf Phytoplasma* (SCWL), ein Stamm der *Rice Yellow Dwarf* Gruppe, und ein Stamm der *Aster Yellows* Gruppe. Dies galt auch für Zuckerrohrpflanzen aus hawaiianischen Plantagen, obwohl bei diesen routinemäßig eine Heißwasser-Behandlung ihrer Setzlinge, welche Phytoplasma eliminieren könnte, durchgeführt wird.

Proben von Zuckerrohrpflanzen anderer Länder (Kuba, Ägypten, Syrien und Thailand), in denen Pflanzen mit Vergilbungs- oder Bleichungssymptomen festgestellt werden, konnten ebenfalls getestet. *Aster Yellows* und *X-Disease* Phytoplasmen fand man in kubanischen Kultivaren, während ein ägyptisches Kultivar *Grassy Shoot Phytoplasma* (ebenfalls ein Stamm der *Rice Yellow Dwarf* Gruppe) enthielt. Zwei andere Kultivare aus Ägypten waren phytoplasmafrei. Auch das syrische Zuckerrohr war von einem Phytoplasma der *Rice Yellow Dwarf* Gruppe infiziert. Unseres Wissens sind das die ersten Nachweise von Phytoplasma in Zuckerrohr aus Hawaii, Ägypten und Syrien. Die Analysen an thailändischen Pflanzen bestätigten publizierte Ergebnisse, dass mit *Sugarcane White Leaf (SCWL) Phytoplasma* infizierte Pflanzen mit White Leaf Disease in Zusammenhang stehen (Nakashima et al., 1994; Wongkaew et al., 1997).

Q-PCR (real-time PCR) gilt als empfindliche, spezifische und rasche Methode um Phytoplasma in Pflanzenmaterial zu messen, dies erwies sich aber offensichtlich nicht für Zuckerrohr mit niedrigem Phytoplasma-Titer. Deshalb wurde nested-PCR als die sensitivere Methode, um Phytoplasma-Infektion niedrigen Titers bei Zuckerrohr festzustellen, angewandt. Ein BLAST-search zeigte, dass die 16S rRNA der gefundenen Phytoplasma-Stämme 99-100% Sequenzidentität mit Phytoplasmen der *Aster Yellows*, *X-Disease* und *Rice Yellow Dwarf* Gruppen aufweisen, sodass keiner davon als neue "*Candidatus* phytoplasma

Art" beschrieben werden kann. Das Phytoplasma aus dem kommerziellen hawaiianischen Kultivar H78-7750 gruppierte sich in *Sugarcane White Leaf Phytoplasma*(SCWLP) ein, zusammen mit einem Stamm aus Taiwan. Es erscheint also möglich, dass über Insekten als Vektoren oder infizierte Setzlinge Phytoplasma aus Taiwan nach Hawaii kam oder umgekehrt. Das thailändische Phytoplasma steht am nächsten dem *White Leaf Phytoplasma* aus Myanmar.

Gewebeschnitte im Transmissions-Elektronenmikroskop (TEM) zeigten, dass Phytoplasma ausschließlich in den Siebröhren der Leitbündel zu finden ist, nicht in Geleitzellen, Phloemparenchym oder anderen Blattzellen. Die normalerweise grünen Gewebe der infizierten Blätter (Bündelscheide und Mesophyll) zeigten starke zytologische Veränderungen wie Akkumulation von Stärkekörnern, eine große Anzahl von Plastoglobuli und desorganisierte Strukturen in Chloroplasten.

Vektoren für Phytoplasma sind Phloemsauger, jedoch wurde bisher keine Blattlaus als Vektor nachgewiesen. Es konnte gezeigt werden, dass die schwarze Zuckerrohrlaus *Melanaphis sacchari*, die der wichtigste Vektor für *Sugarcane Yellow Leaf Virus* ist, Phytoplasma nicht übertragen kann. Heißwasser-Behandlung war als Methode zum Abtöten von Phytoplasma in Pflanzenteilen beschrieben worden. Dies konnte bestätigt werden und eine 60-minütige Behandlung in 50° heißem Wasser kann für die Eliminierung von Phytoplasma in Zuckerrohrsetzlingen empfohlen werden.

7. Acknowledgement

First of all, I would like to acknowledge my supervisor Prof. Dr. Ewald Komor, who gave me the acceptance to come to University of Bayreuth and to work in his lab.

Thanks to all people in my department (plant physiology) for their help, especially Prof. Dr. Stephan Clemens, Christiane Meinen and Ursula Ferrera.

I would like to thank Dr. Eric Hummel, Daniel Souza, Christian Seybold, Philipp Gasch and Thomas Liebenstein; they are not only colleagues but nice friends who kindly helped me with all kinds of situations.

I appreciate the assistance from Dr. Alfons Weig to do phylogenetic analysis.

I am grateful to Rita Grotjahn (technician in the laboratory of electron microscopy) for preparation of all ultrathin sections.

This is the right place to thank Prof. Dr. A. Bertaccini at University of Bologna, Italy; Prof. Dr. E. Seemüller at institute of plant protection in Dossenheim, Germany and Dr. J. Hodgetts at University of Nottingham, UK for their cooperation and advices throughout my working process.

My scholarship was funded by University of Aleppo, Syria. I am very thankful for this support.

I would like to thank my supervisor Dr. Nada Omlah at University of Aleppo, Syria.

I am extremely thankful to my parents for their love and care which gives me the greatest motivation in life and for their encouragement with every step I take.

And God,
I am forever very thankful for being with me all the time and help me with your best blessings.

8. References

Abu Ahmad, Y., Rassably, L., Royer, M., Borg, Z., Braithwaite, K. S., Mirkov, T. E., Irey, M.S., Perrier, X., Smith, G.R., Rott, P. (2006). Yellow leaf of sugarcane is caused by at least three different genotypes of sugarcane yellow leaf virus, one of which predominates on the Island of Réunion. *Archive of Virology* **151**, 1355-1371.

Agrios, G. N. (1997). Plant pathology (4th ed. Academic Press, San Diego, USA).

Agrios, G. N. (2004). Plant pathology (5th ed. Academic Press, San Diego, USA).

Aljanabi, S. M., Parmessur, Y., Moutia, Y., Saumtally, S., Dookun, A. (2001). Further evidence of the association of a phytoplasma and a virus with yellow leaf syndrome in sugarcane.*Plant Pathology* **50**, 628-636.

Ammar, M. I., Amer, M. A., Rashed, M. F. (2005). Detection of phytoplasma associated with yellow streak disease of date palms (*Phoenix dactylifera* L.) in Egypt.*Egyptian Journal of Virology* **2**, 74-86.

Ammar, E. D., Hogenhout, S. A. (2006). Mollicutes associated with arthropods and plants (CRC Press, Boca Raton, FL, USA).

Andersen, M., Beaver, R., Gilman, A., Liefting, L., Balmori, E., Beck, D., Sutherland, P., Bryan, G., Gardner, R., Forster, R. (1998). Detection of phormium yellow leaf phytoplasma in New Zealand flax (*Phormium tenax*) using nested PCRs. *Plant Pathology* **47**, 188-196.

Ariyarathna, H., Everard, J., Karunanayake, E. H. (2007). Diseased sugarcane in Sri Lanka is infected with sugarcane grassy shoot and/or sugarcane white leaf phytoplasma. *Australasian Plant Disease Notes* **2**, 123-125.

Arocha, Y., Gonzales. L., Peralta, E. L., Jones, P. (1999). First report of virus and phytoplasma pathogens associated with yellow leaf syndrome of sugarcane in Cuba. *Plant Disease* **83**, 1171.

Arocha, Y., Jones, P., Sumac, I., Peralta, E. L. (2000). Detection of phytoplasmas associated with yellow leaf syndrome in Cuba. *Revista de Protección Vegetal* **15**, 81-87.

References

Arocha, Y., Lopez, M., Piñol, B., Fernandez, M., Picornell, B., Almeida, R., Palenzuela, I., Wilson, M.R., Jones, P. (2005a). "Candidatus Phytoplasma graminis" and "candidatus Phytoplasma caricae", two novel phytoplasmas associated with diseases of sugarcane, weeds and papaya in Cuba. *International Journal of Systematics and Evolutionary Microbiology* **55**, 2451-2463.

Arocha, Y., Lopez, M., Fernandez, M., Pinol, B., Horta, D., Peralta, E. L., Almeida, R., Carvajal, O, Picornell, S., Wilson, M.R., Jones, P. (2005b).Transmission of a sugarcane yellow leaf phytoplasma by the delphacid plant hopper Saccharosydne saccharivora, a new vector of sugarcane yellow leaf syndrome.*Plant Pathology* **54**, 634-642.

Asche, M. (1997). A review of the systematics of Hawaiian planthoppers (Hemiptera: Fulguroidea).*Pacific Science* **51**, 366-376.

Aslam, K. (2001). Plant diseases (Gyan Publishing House, New Delhi, India).

Bai, X., Zhang, J., Ewing, A., Miller, S.A., Radek, A.,Schevchenko D., Tsukerman, K., Walunas, T., Lapidus,A., Campbell, J. W., Hogenhout, S. A. (2006). Living with genome instability: the adaptation of phytoplasmas to diverseenvironments of their insect and plant hosts. *Journalof Bacteriology* **188**, 3682-3696.

Bailey, R. A., Bechet, G. R., Cronje, C. R. (1996). Notes on the occurrence of yellow leaf syndrome of sugarcane in southern Africa. *South African Sugar Technology Association Proceedings* **70**, 3-6.

Bertamini, M., Grando, M.S., Muthuchelian, K., Nedunchezhian, N. (2002a). Effect of phytoplasmal infection on photosystem II efficiency and thylakoid membrane protein changes in field grown apple (*Maluspumila*) leaves. *Physiological and Molecular Plant Pathology* **61**, 349-356.

Bertamini, M., Nedunchezhian, N., Tomasi, F., Grando, S. (2002b). Phytoplasma (Stolbur subgroup (Bois Noir-BN)) infection inhibits photosynthetic pigmentsribulose-1,5-biphosphate carboxylase and photosynthetic activities in field grown grapevine (*Vitis vinifera* L. cv. Chardonnay) leaves. *Physiological and Molecular Plant Pathology* **61**, 357-366.

Bianco, P. A., Davis, R. E., Prince, J. P., Lee, I-M., Gundersen, D. E., Fortusini, A., Belli, G. (1993). Double and single infections by ester yellows and elm yellow MLOs in grapevines with symptoms characteristic of flavescence doree. *Rivistadi Patologia Vegetale* **3**, 69-82.

Biswal, U. C, Biswal, B. (1988). Ultrastructural modifications and biochemical changes during senescence of chloroplasts.*International Review of Cytology* **113**, 271-321.

Borth, W., Hu, J. S., Schenk, S. (1994).Double-stranded RNA associated with sugarcane yellow leaf syndrome. *Sugar Cane* **3**, 5-8.

Borth, W. B., Hu, J. S., Kirkpatrick, B. C., Gardner, D. E., German, T.L. (1995). Occurrence of phytoplasmas in Hawaii.*Plant Disease* **79**, 1094-1097.

Borth, W. B, Fukuda, S. K., Harmasaki, R.T., Hu, J. S., Almeida, R. P. (2006). Detection, characterisation and transmission by Macrosteles leaf hoppers of Water cress yellows phytoplasma in Hawaii. *Annals of Applied Biology* **149**, 357-363.

Borth, W. B., Hamasaki, R. T., Ogata, D., Fukuda, S. K., Hu, J. S. (2002).First report of phytoplasma infecting water cress in Hawaii.*Plant Disease* **86**, 331.

Cainelli, C., Forno, F., Mattedi, L., Grando, M. S. (2007). Can apple aphids be vectors of "Candidatus phytoplasma mali"? *IOBC/WPRS Bulletin* **30**, 261-266.

Caudwell, A., Larrue, J., Boudon, Padieu. E., McLean, G. D. (1997). Flavescence dorée elimination from dormant wood of grapevines by hot-water treatment.*Australian Journal of Grape and Wine Research* **3**, 21-25.

Chalak, L., Elbitar, A., Rizk, R., Choueiri, E., Salar, P., Bovej, M. (2005). Attempts to eliminate '*Candidatus* phytoplasma phoenicium' from infected Lebanese almond varieties by tissue culture techniques combined or not with thermo therapy. *European Journal of Plant Pathology* **112**, 85-89.

Chen, C. T. (1974). Sugarcane white leaf disease in Thailand and Taiwan. *Sugarcane Pathology News* **11**, 12-23.

Chen, C. T., Kusalwong, A. (2000). White leaf. In: Rott, P., Bailey, R. A., Comstock, J. C., Croft, B. J., Saumtally, A. S., eds. *A guide to sugarcane diseases*. Montpellier, CIRAD and ISSCT, pp. 231-236.

Cheung, W. Y., Hubert, N., Landry, B. S. (1993). A simple and rapid DNA microextraction method for plant, animal, and insect suitable for RAPD and other PCR analyses.*Genome Research* **3**, 6970.

Chona, B. L. (1958). Some diseases of sugarcane reported from India in recent years. *Indian Phytopathology* **11**, 1-9.

Chona, B. L., Capoor, S. P., Varma, P. M., Seth, M. L. (1960). Grassy shoot disease of sugarcane. *Indian Phytopathology* **13**, 37-47.

Christensen, N. M., Nicolaisen, M., Hansen, M., Schulz, A. (2004). Distribution of phytoplasmas in infected plants as revealedby real-time PCR and bioimaging. *Molecular Plant-Microbe Interaction* **17**, 1175-1184.

Christensen, N. M., Axelsen, K. B., Nicolaisen, M., Schulz, A. (2005). Phytoplasmas and their interactions with hosts.*Trends in Plant Science* **10**,526-535.

Ciancio, A., Mukerji, K. G. (2008). Integrated management of diseases caused by fungi, phytoplasma and bacteria(Springer Science and BusinessMedia B.V.).

Comstock, J. C., Irvine, J. E., Miller, J. D. (1994). Yellow leaf syndrome appears on the United States mainland. *Sugar Journal* **56**, 33-35.

Contaldo, N., Soufi, Z., Bertaccini, A. (2011). Preliminary identification of phytoplasmas associated with grapevine yellows in Syria. *Bulletin of Insectology* **64**, 217-218.

Cronje, C. R., Tymon, A. M., Jones, P., Bailey, R. A. (1998). Association of a phytoplasma with a yellow leaf syndrome of sugarcane in Africa.*Annals of Applied Biology* **133**, 177-186.

Dai, Q., He, F. T., Liu, P. Y. (1997). Elimination of phytoplasma by stem culture from mulberry plants (*Morus alba*) with dwarf disease. *Plant Pathology* **46**, 56-61.

Davies, R. E., Whitcomb, R. F., Steere, R. L. (1968). Remission of aster yellows disease by antibiotics. *Science* **161**,793-794.

Davies, D. L, Clark, M. F. (1992). Production and characterization of polyclonal and monoclonal antibodies against peach yellow leaf roll MLO-associated antigens. *Acta Horticulturae* **309**, 275-283.

Davis, R. E., Sinclair, W. A. (1998). Phytoplasma identity and disease etiology.*Phytopathology* **88**, 1372-1376.

Deng, S., Hiruki, C. (1991). Amplification of 16S rRNA genes from culturable and nonculturable Mollicutes.*Journal of Microbiological Methods* **14**, 53-61.

References

Doi, Y., Teranaka, M., Yora, K., Asuyama, H. (1967). Mycoplasma- or PLT group-like microorganisms found in the phloem elements of plants infected with mulberry dwarf, potato witches' broom, aster yellows or paulownias witches broom. *Annals of the Phytopathologicial Society of Japan* **33**, 259-266.

Donald, M. C.(2006). Contributing Editor (Chapter 15) Harvard Medical school current protocols in molecular biology.15.0.1-15.0.3.

Doyle, J., Doyle, J. (1990). Isolation of plant DNA from fresh tissue.*Focus* **12**, 13-15.

Edgar, R. C. (2004). MUSCLE: a multiple sequence alignment method with high accuracy and high throughput. *Nucleic Acids Research* **32**, 1792-1797.

Edison, S., Ramakrishnan, K. (1972). Aerated steam therapy for the control of grassy shoot disease (GSD) of sugarcane. *Mysore Journal of Agricultural Sciences* **6**, 492-494.

Edison, S., Ramakrishnan, K., Narayanasamy, P. (1976).Comparison of grassy shoot disease (India) with the white leaf disease (Taiwan) of sugarcane.*Sugarcane Pathology News* **17**, 30-35.

Esau, K. (1965). Plant Anatomy (2nd ed. John Wiley, New York).

Firrao, G., Smart, C. D., Kirkpatrick, B. C. (1996). Physical map of the western X disease phytoplasma chromosome. *Journal of Bacteriology* **178**, 3985-3988.

Firrao, G., Gibb, K., Streten, C. (2005). Short taxonomic guide to the genus '*Candidatus phytoplasma*'. *Journal of Plant Pathology* **87**, 249-263.

Franova,J. (2011). Difficulties with conventional phytoplasma diagnostic using PCR/RFLP analyses. *Bulletin of Insectology* **64**,287-288.

Gaur, R.K., Raizada,R., Rao,G.P. (2008). Sugarcane yellow leaf phytoplasma associated for the first time with sugarcane yellow leaf syndrome in India. *Plant Pathology* **57**, 772-772.

Goodwin, P. H., Xue, B. G., Kuske, C.R., Sears, M. K. (1994). Amplification of plasmid DNA to detect plant pathogenic mycoplasma like organisms.*Annals of Applied Biology* **124**, 27-36.

References

Govind, P. R., Sangeeta, S., Maneesha, S., Carmine, M. (2007). Phylogenetic relationships of sugarcane grassy shootphytoplasma with closely related agents. *Bulletin of Insectology* **60**,347-348.

Grisham, M. P., Pan, Y. B., White, W. H. (2002). Potential effect of yellow leaf syndrome on the Louisiana sugarcane industry. *Journal of the American Society of Sugar Cane Technologists* **22**, 125-126.

Guindon, S., Gascuel, O. (2003).A simple, fast and accurate algorithm to estimate large phylogenies by maximum likelihood. *Systematic Biology* **52**, 696-704.

Gundersen, D. E., Lee, I-M., Rehner, S. A., Davis, R. E., Kingsbury, D. T. (1994). Phylogeny of mycoplasmalike organisms (phytoplasmas): a basis for their classification. *Journal of Bacteriology* **176**, 5244-5254.

Hanboonsong, Y., Panyim, S., Damak, S. (2002). Transovarial transmission of sugarcanewhite leaf phytoplasma in the insect vector *Matsumuratettix hiroglyphicus* (Matsumura).*Insect Molecular Biology* **11**, 97-103.

Harrison, S. D., Broadie, K., van de Goor, J., Rubin, G. M. (1994). Mutations in the Drosophila Rop gene suggest a function in general secretion and synaptic transmission. *Neuron* **13**, 555-566.

Harrison, N. A. (1999).Phytoplasma Taxonomy. First internet conference on phytopathogenic mollicutes.

Harrison, N. A., Womack, M., Carpio, M. L.(2002). Detection and characterization of a lethal yellowing (16SrIV) group phytoplasma in Canary Island date palms affected by lethal decline in Texas. *Plant Disease* **86**,676-681.

Heinrich, M., Botti, S., Caprara, L., Arthofer, W., Strommer, S., Hanzer, V., Katinger, H., Bertaccini, A., Laimer, M. (2001). Improved detection methods for fruit tree phytoplasmas. *Plant Molecular Biology Reporter* **19**, 169-179.

Herson, J. M., French, R. (1993). The polymerase chain reaction and plant disease diagnosis.*Annual Review of Phytopathology* **31**, 81-109.

References

Hodgetts, J., Ball, T., Boonham, N., Mumford, R., Dickinson, M. (2007). Use of terminal restriction fragment length olymorphism (T-RFLP) for identification of phytoplasmas in plants.*Plant Pathology* **56**, 357-365.

Hodgetts, J., Dickinson, M. (2010). Phytoplasma phylogeny and detection based on genes other than 16S rRNA. In: Weintraub, P. G., Jones, P, eds. *Phytoplasmas Genomes, Plant Hosts and Vectors*. Wallingford, UK, CAB International, pp. 93-113.

Jones, P.(2002). Phytoplasma plant pathogens (CABI Publishing,Wallingford, UK).

Jung, H.Y., Sawayanagi, T., Wongkaew, P., Kakizawa, S., Nishigawa, H., Wei, W., Oshima, K., Miyata, S-I., Ugaki, M., Hibi, T., Namba, S. (2003). '*Candidatus* Phytoplasma oryzae', a novel phytoplasma taxon associated with rice yellow dwarf disease. *International Journal of Systematic and Evolutionary Microbiology* **53**, 1925-1929.

Junqueira, A., Bedendo, I., Pascholati, S. (2004). Biochemical changes in corn plants infected by the maize bushy stunt phytoplasma. *Physiological and Molecular Plant Pathology* **65**, 181-185.

Kirkpatrick, B. C., Fisher, G. A., Fraser, J. D., Purcell, A. H. (1990). Epidemiological and phylogenetic studies on western X-disease mycoplasma-like organisms. *International Journal of Medical Microbiology* **20**, 288-297.

Kirkpatrick, B. C. (1992). Mycoplasma-like organisms (2nd ed. Springer, New York, USA).

Kirkpatrick, B. C., Smart, C. D., Gardner, S. L., Gao, L., Ahrens, U., Maurer, R., Schneider, B., Lorenz, H., Seemüller, E. (1994). Phylogenetic relationships of plant pathogenic MLOs established by 16/23S rDNA spacer sequences. *IOM Letters* **3**, 228-229.

Komor, E. (2011). Susceptibility of sugarcane, plantation weeds and grain cereals to infection by sugarcane yellow leaf virus and selection by sugarcane breeding in Hawaii. *European Journal of Plant Pathology* **129**, 379-388.

Komor, E., El-Sayed, A., Lehrer, A. T. (2010). Sugarcane yellow leaf virus introduction and spread in Hawaiian sugarcane industry: Retrospective epidemiological study of an unnoticed, mostly asymptomatic plant disease. *European Journal of Plant Pathology* **127**, 207-217.

Lederer, W., Seemuller, E. (1991). Occurrence of mycoplasma-like organisms in diseased and non-symptomatic alder trees (Alnus spp). *European Journal of Forest Pathology* **21**, 90-96.

Lee, I. M.(1999). Molecular-based methods for the detection and identification of phytoplasmas. First internet conference on phytopathogenic mollicutes.

Lee, I. M., Davis, R. E. (1992).Mycoplasmas which infect plant and insects. In: Molecular Biology and Pathogenesis (Maniloff J., McElhansey, R.N., Finch, L.R. and Baseman, J.B., eds).*Washington, USA,American Society of Microbiology,*379-390.

Lee, I. M., Hammond, R. W., Davis, R. E., Gunderson, G. E. (1993). Universal amplification and analysis of pathogen 16S rDNA for classification and identification of mycoplasma like organism. *Phytopathology* **83**, 834-832.

Lee, I. M., Gundersen-Rindal, D. E., Davis, R. E, Bartoszyk, I. M. (1998). Revised classification scheme of phytoplasmas based on RFLP analyses of 16S rRNA and ribosomal protein gene sequences. *International Journal of Systematic Bacteriology* **48**, 1153-1169.

Lee, I. M., Gundersen-Rindal, D. E., Davis, R. E.(2000). Phytoplasma: Phytopathogenic Mollicutes. *Annual Review of Microbiology* **54**, 221-255.

Lehrer, A. T., Schenck, S., Yan, S. L., Komor, E. (2007). Movement of aphid-transmitted Sugarcane yellow leaf virus (ScYLV) within and between sugarcane plants. *Plant Pathology* **56**, 711-717.

Lehrer, A. T., Komor, E. (2008). Symptom expression of yellow leaf disease in sugarcane cultivars with different degrees of infection by Sugarcane yellow leaf virus. *Plant Pathology* **57**, 178-189.

Lehrer, A. T., Wu, K.K., Komor, E. (2009). Impact of Sugarcane yellow leaf virus (SCYLV) on growth and sugar yield of sugarcane. *Journal of General Plant Pathology* **75**, 288-296.

Lepka, P., Stitt, M., Moll, E., Seemuller, E. (1999). Effect of phytoplasmal infection on concentration and translocation of carbohydrates and amino acids in periwinkle and tobacco. *Physiological and Molecular Plant Pathology* **55**, 59-68.

Lherminier, J., Prensier, G., Boudon-Padieu, E., Caudwell, A. (1990). Immunolabeling of grapevine flavescence doree MLO in salivaryglands of Euscelidius variegatus: a light and electron microscopy study. *Journal of Histochemistry and Cytochemistry* **38**, 79-85.

Lim, P. O., Sears, B. B. (1989). 16S rRNA sequence indicates that plant-pathogenic mycoplasmalike organisms are evolutionarily distinct from animal mycoplasmas. *Journalof Bacteriology* **171**, 5901-5906.

Lindqvist, R. (1999). Detection of *Shigella* spp. in food with a nested PCR method-sensitivity and performance compared with a conventional culture method. *Journal of Applied Microbiology* **86**,971-978.

Ling, K.C. (1962). White leaf disease of sugarcane. *Taiwan Sugar* **9**, 1-5.

Lockhart, B. E., Cronje, P. R. (2000). Yellow leaf syndrome. In:Rott, P., Bailey, R. A., Comstock, J. C., Croft, B. J., Saumtally, A. S, eds.*A guide to sugarcane diseases.*Montpellier, CIRAD and ISSCT, pp. 291-295.

Loomis, W. D. (1974).Overcoming problems of phenolic and quinines in the isolation ofplant enzymes and organelles. *Methods in Enzymology* **31**, 528-545.

Mangelsdorf, A. J. (1962). A research program for the Thailand sugar industry. Bangkok: department of agriculture.

Marcone, C. (2002). Phytoplasma diseases of sugarcane. *Sugarcane Technology* **4**, 79-85.

Marcone, C., Schneider, B., Seemuller, E. (2004). '*Candidatus* Phytoplasma cynodontis', the phytoplasma associated with Bermuda grass white leaf disease. *International Journal of Systematic and Evolutionary Microbiology* **54**, 1077-1082.

Matsumoto, T., Lee, C.S., Teng, W.S. (1968). Studies on sugarcane white leaf disease of Taiwan, with special reference to transmission by a leafhopper, *Sugarcane Technology* **13**, 1090-1098.

McCoy, R. E., Caudwell, A., Chang, C. J., Chen, T.A., Chiykowski, L. N., Cousin, M. T., Dale, J. L., DeLeeuw, G. T. N., Golino, D. A., Hackett, K. J., Kirkpatrick, B. C., Marwithz, R., Petzold, H., Sinha, R. C., Sugiura, M., Whitcomb, R. F., Yang, I. L., Zhu, B. M., Seemuller, E. (1989). Plant diseases associated with mycoplasma-like organisms (Academic Pres, New York, USA).

Mishra, S.R. (2004). Mollicutes and plant diseases (Discovery Publishing House, New Delhi, India).

Moonan, F., Molina, J., Mirkov, T. E. (2000). Sugarcane yellow leaf virus: an emerging virus that has evolved by recombination between luteoviral and poleroviral ancestors. *Virology* **269**, 156-171.

Murral, J. D., Nault, L. R., Hoy, C. W., Madden, L. V., Miller, S. A. (1996). Effects of temperature and vector age on transmission of two Ohio strains of aster yellows phytoplasma by the aster leafhopper (*Homoptera: Cicadellidae*). *Journal of Economic Entomology* **89**, 1223-1232.

Musetti, R. (2006). Patogeni e piante di interesse agronomico: un approccio morfologico. In: Quaglino, D., Falcieri, E., Catalano, M., Diaspro, A., Montone, A., Mengucci,P., Pellicciari, C, eds.*1956–2006: 50 anni di Microscopia in Italia trastoria, progresso ed evoluzione.* PI.ME Editrice, Pavia, Italy, pp. 325-334.

Nakashima, K., Chaleeprom, W., Wongkaew, P., Sirithorn, P.(1994).Detection of mycoplasma-like organisms associated withwhite leaf disease of sugarcane in Thailand using DNAprobes. *Japan International Research Center for AgriculturalSciences* **1**, 57-67.

Nakashima, K., Hayashi, T., Chaleeprom, W., Wongkaew, P., Sirithorn, P. (1996). Complex phytoplasma flora in Northest Thailand as revealed by 16*s* rDNA analysis. *Annals of the Phytopathological Society of Japan* **62**, 57-60.

Namba, S., Oyaizu, H., Kato, S., Iwanami, S., Tsuchizaki, T. (1993). Phylogenetic diversity of phytopathogenic mycoplasmalike organisms. *International Journal of Systematic and Evolutionary Microbiology* **43**, 461-467.

Noode´n, L. D. (1988). The phenomena of senescence and aging (Academic Press, San Diego, USA).

Omar, F. A., Emeran, A. A., Abass, M. J. (2008). Detection of phytoplasma associated with periwinkle virescence in Egypt. *Plant Pathology Journal,* **7**, 92-97.

Oshima, K., Kakizawa, S., Nishigawa, H., Jung, H.Y., Wei, W., Suzuki, S., Arashida, R., Nakata, D., Miyata, S., Ugaki, M., Namba, S. (2004). Reductive evolution suggested from the complete genome sequence of a plant-pathogenic phytoplasma.*Nature Genetics* **36**, 27-29.

References

Parmessur, Y., Aljanabi, S., Saumtally, S., Dookunsaumtally, A. (2002). *Sugarcane yellow leaf virus* and sugarcane yellows phytoplasma: elimination by tissue culture. *Plant Pathology*, **51**, 561-566.

Pérez de Rozas, A.M., González, J., Aloy, N., Badiola, I. (2008). Standardization of nested-PCR for the detection of *Pasteurella multocida, Staphylococcus aureus*, myxomatosis virus, and rabbit Haemorrhagic disease virus.*Pathology and Hygiene* 9th World Rabbit Congress – June 10-13, 2008 – Verona – Italy.

Purcell, A. H. (1985). The ecology of bacterial and mycoplasma plant diseases spread by leafhoppers and planthoppers.In:Nault,L. R., Rodriguez, J. G, eds. *The Leafhoppers and Planthoppers.*JohnWiley and Sons, New York, NY, USA, pp. 351-380.

Purves, W. K., Orians, G. H., Heller, H. C.(1992). Life: The science of biology (3rdedition. Sinauer Associates, Sunderland, Mass,UK).

Rao, G. P., Singh, A., Singh, H. B., Sharma, S. R. (2005). Phytoplasma diseases of sugarcane: characterization, diagnosis and management. *Indian Journal of Plant Pathology* **23**, 1-21.

Rao, G. P., Srivastava, S., Singh, M., Marcone, C. (2007).Phylogenetic relationships of sugarcane grassy shoot phytoplasma with closely related agents. *Bulletin of Insectology* **60**, 347-348.

Ratana, S. (2001).Recent studies on white leaf and grassy shoot phytoplasma of sugarcane. In: Rao, G. P., Ford, R. E., Tosic, M., Teakle, D. S, eds. *Sugarcane Pathology*, Vol. II: Virus and Phytoplasma Disease. Enfield, NH, USA: Science Publishers Inc, pp. 235-244.

Rishi, N., Chen, C. T. (1989). Grassy shoot and white leaf disease.In:Ricaus, B. C., Egan, B. T, eds. *Diseases of Sugarcane.*Elsevier Science Publisher, Amsterdam,pp. 289-300.

Roberts, R. J., Kenneth, M. (1976).Restriction endonucleases.*CRC CriticalReviews in Biochemistry*.**4**, 123-164.

Rogers, P. F. (1969). Proceedings of a Meeting on the Yellow Wilt Condition of Sugarcane.June 25th–26th 1969. Nairobi, Kenya: *East African Specialist Committee on Sugarcane Research.*

References

Rutherford, R. S., Brune, A. E., Nuss, K. J. (2004). Current status of research onsugarcane yellow leaf syndrome in southern Africa. *Proceedings. Congress of the South African Sugar Technologists Association* **78**, 173-180.

Sarindu, N., Clark, M. F. (1993). Antibody production and identity of MLOs associated with sugar-cane white leaf diseases and bermuda-grass white leaf disease from Thailand. *Plant Pathology* **42**, 396-402.

Schenck, S. (1990).Yellow leaf syndrome – a new sugarcane disease.*Hawaiian Sugar Planters Association: Annual Report* 38-39.

Schenck, S., Hu, J., Lockhart, B. (1997). Use of a tissue blot immunoassay to determine the distribution of sugarcane yellow leaf virus in Hawaii. *Sugar Cane* **4**, 5-8.

Schenck, S., Lehrer, A. T. (2000). Factors affecting the transmission and spread of Sugarcane Yellow Leaf Virus. *Plant Disease* **84**, 1085-1088.

Schneider, B., Ahrens, U., Kirkpatrick, B. C., Seemüller, E. (1993). Classification of plant-pathogenic mycoplasma-like organisms using restriction-site analysis of PCR-amplified 16S rDNA. *Journal of General Microbiology* **139**, 519-527.

Schneider, B., Gibb, K. S. (1997). Detection of phytoplasmas in declining pears in southern Australia. *Plant Disease* **81**, 2548.

Sdoodee, R., Schneider, B., Padovan, A.C. and Gibb, K.S. (1999). Detection and gentic relatedness of phytoplasma associated with plant diseases in Thailand. *The Journal of Biochemistry, Molecular Biology and Biophysics* **3**, 133-140.

Seemüller, E. (1976). Investigations to demonstrate mycoplasma-like organisms in diseased plants by fluorescence microscopy.*ActaHorticulturae* **67**, 109-112.

Seemüller, E., Schneider, B., Maurer, R., Ahrens, U., Daire, X., Kison, H., Lorenz, K.H., Firrao, G., Avinent, L., Sears, B.B. (1994).Phylogenetic classification of phytopathogenic mollicutes by sequence analysis of 16S ribosomal DNA. *International Journal of Systematic Bacteriology* **44**, 440-446.

Siller, W. et al. (1987). Occurrence of mycoplasma-like organisms in parenchyma cells of Cuscuta odorata (Ruiz-Et-Pav). *Journal of Phytopathology* **119**, 147-159.

References

Sinclair, W. A., Griffiths, H. M., Lee, I. M. (1994).Mycoplasmalike organisms as causes of slowgrowth and decline of trees and shrubs. *Journal of Arboriculture* **20**, 176-189.

Skrzeczkowski, L. J., Howell, W. E., Eastwell, K. C., Cavileer, T. D. (2001). Bacterial sequences interferring in detection of phytoplasma by PCR using primers derived from the ribosomal RNA operon. *Acta Horticulturae* **550**, 417-424.

Smart, C. D., Schneider, B., Blomquist, C. L., Guerra, L. J., Harrison, N. A., Ahrens, U., Lorenz, K. H., Seemüller, E., Kirkpatrick, B. C. (1996). Phytoplasma-specific primers based on sequences of the 16S-23S rRNA spacer region. *Applied and Environmental Microbiology* **62**, 2988-2993.

Smith, G. R., Borg, Z., Lockhart, B. L., Braithwaite, K. S., Gibbs, M. J. (2000). Sugarcane yellow leaf virus: a novel member of the Luteoviridae that probably arose by interspecies recombination. *Journal of General Virology* **81**, 1865-1869.

Snounou, G., Viriyakosol, S., Zhu, X. P, Jarra, W., Pinheiro, L., do Rosario, V. E., Thaithong, S., Brown, K. N. (1993). High sensitivity of detection of human malaria parasites by the use of nested polymerase chain reaction. *Molecular and Biochemical Parasitology* **61**, 315-320.

Spurr, A. R.(1969). A low-viscosity epoxy resin embedding medium for electron microscopy. *Ultrastructure Research* **26**, 31-43.

Srivastava, S., Singh, V., Gupta, P. S., Sinha, O. K. (2003).Detection of phytoplasma of GSD of sugarcane based on PCR assay using ribosomal RNA sequences. In: Singh, V., Sinha, O, K, eds. *National Seminar on Emerging Trends in Plant Disease Research and Management.*Lucknow, India: Indian Institute of Sugarcane Research, 37.

Srivastava, S., Singh, V., Gupta, P. S., Sinha, O. K., Batiha, A. (2005). Nested PCR assay for detection of sugarcane grassy shoot phytoplasma in the leafhopper vector Deltocephalus vulgaris: a first report. *Plant Pathology* **55**, 25-28.

Suma, S., Jones, P. (2000).Ramu stunt. In: Rott, P., Bailey, R.A., Comstock, J.C., Croft, B. J., Saumtally, A. S, eds. *A guide to sugarcane diseases*. Montpellier, CIRAD and ISSCT, pp. 226-230.

References

Suzuki, S., Oshima, K., Kakizawa, S., Arashida, R., Jung, H.Y., Yamaji, Y., Nishigawa, H., Ugaki, M., Namba, S. (2006). Interactionbetween the membrane protein of a pathogen and insect microfilamentcomplex determines insect-vector specificity. *Proceedings of the National Academy of Sciences of the United States of America* **103**, 4252-4257.

Tassart-Sublirats, V., Clair, D., Grenan, S., Boudon-Padieu E., Larrue, J. (2003). Hot water treatment: curing efficiency for phytoplasmas infection and effect on plant multiplication material. In: *Extended Abstracts 14° ICVG Conference*, Locorotondo (BA), Italy, pp. 69-70.

Thomson, W. W., Platt-Aloia, K. A. (1987).Ultrastructure and senescence in plants.Plant senescence: its biochemistry and physiology, Rockville, Maryland).

Tran-Nguyen, L., Blanche, K. R., Egan, B, Gibb, K. S. (2000). Diversity of phytoplasmas in Northern Australian sugarcane and other grasses.*Plant Pathology* **49**, 666-669.

Uyemoto, J. K., Connell, J. H., Hasey, J. K., Luhn, C. F. (1992). Almond brown line and decline: a new disease probably caused by a mycoplasma-like organism. *Annals of Applied Biology* **120**, 417-24.

Valiunas, D., Urbanaviciene, L., Jomantiene, R., Davis, R. E. (2007). Molecular detection, classification and phylogenetic analysis of subgroub 16SrI-C phytoplasmas detected in diseased Poa and Festuca in Lithuania. *Biologija* **53**, 36-39.

Vega, J., Scagliusi, M. M., Ulian, E. C. (1997). Sugarcane yellow leaf disease in Brazil: evidence of association with a luteovirus. *Plant Disease* **81**, 21-26.

Viswanathan, R. (2000). Grassy shoot. In: Rott, P., Bailey, R. A., Comstock, J. C., Croft, B. J., Saumtally, A. S, eds. *A guide to sugarcane diseases*. Montpellier, CIRAD and ISSCT, pp. 215-220.

Weintraub, G. P., Beanland, L. (2006). Insect vector of phytoplasmas.*Annual Review of Entomology* **51**, 91-111.

Weintraub, P. G., Jones, P. (2010). Phytoplasmas Genome, Plant Hosts and Vectors (First ed. CABI, UK).

Welliver, R. (1999). Diseases caused by phytoplasmas. *Plant Pathology Circular* No. **82**.

Whitcomb, R. F., Tully, E. D. (1989). The Mycoplasmas (Academic Press, Inc, San Diego, USA).

Wongkaew, P. (1999). Sugarcane white leaf disease and control strategies (Thailand Research Fund. T and R Celeca, Bangkok).

Wongkaew, P., Hanboonsong, Y., Sirithorn, P., Choosal, C., Boonkrong, S., Tinnangwatanna, T., Kitchareonpanya, R., Darnak, S. (1997). Differentiation of phytoplasmas associated with sugarcane and gramineous weed leaf disease and sugarcane grassy shoot disease by RFLP and sequencing. *Theoretical and Applied Genetics* **95**, 660-663.

Yan, S. L., Lehrer, A. T., Hajirezaei, M. R., Springer, A., Komor, E. (2009). Modulation of carbohydrate metabolism and chloroplast structure in sugarcane leaves which were infected by *Sugarcane yellow leaf virus* (SCYLV). *Physiological and Molecular Plant Pathology* **73**, 78-87.

Zhu, Y. J., Lim, T. S., Schenck, S., Arcinas, A., Komor, E. (2010). RT-PCR and quantitative real-time RT-PCR detection of Sugarcane Yellow Leaf Virus (SCYLV) in symptomatic and asymptomatic plants of Hawaiian sugarcane cultivars and the correlation of SCYLV titre to yield. *European Journal of Plant Pathology* **127**, 263-273.

9. List of figures

1. Introduction   1

Figure.1.1. Comparison of sizes of some eubacteria.   3

Figure.1.2. Pleomorphic phytoplasmas in sieve tubes.   4

Figure.1.3. Phytoplasmas and their diseases are worldwide.   6

Figure.1.4. Host cycle of phytoplasmas.   7

Figure.1.5. Phytoplasmas are firmicutes.   8

Figure.1.6. Diagram of the longitudinal view of phloem cells.   11

Figure.1.7. Sugarcane yellow leaf syndrome (YLS).   12

Figure.1.8. Sugarcane white leaf (SCWL).   13

2. Material and Methods   17

Figure.2.1. A diagram illustrating of the method of nested PCR.   25

Figure.2.2. Diagrammatic representation of location of used primer pairs and expected size of their amplified products based on phytoplasma rRNA operon.   29

Figure.2.3. Diagrammatic representation of a phytoplasma rRNA operon and genomic location of primers used for phytoplasma detection.   29

Figure.2.4. Diagram illustrating of SYBR Green during PCR amplification.   35

Figure.2.5. Diagram illustrating of TaqMan probe chemistry mechanism.   36

Figure.2.6. Diagrammatic representation of genomic location of qPCR primers and probe used for phytoplasma detection.   37

3. Results   39

Figure.3.1. Nested PCR-products of positive and negative controls and of a positive control, which was mixed with increasing amounts of sugarcane DNA.   40

Figure.3.2. Phytoplasma in Hawaiian and Cuban sugarcane cultivars.   43

Figure.3.3. Phytoplasma in Egyptian and Syrian sugarcane cultivars.   43

Figure.3.4. Restriction fragment analysis of PCR products from Hawaiian and Cuban sugarcane cultivars containing phytoplasma.   45

Figure.3.5. Nested-PCR assay (II) products (1.2kb) amplified with primers (SN910601/P6, R16F2n/R16R2).   47

Figure.3.6. RFLP profiles of nested-PCR assay (II) products.   48

Figure.3.7. Nested-PCR assay (III) products (0.2kb) amplified with primer pair (MLO-X/MLO-Y, P1/P2).   49

Figure.3.8. Nested-PCR assay (IV) products.   50

List of figures

Figure.3.9. RFLP profiles of nested-PCR assay (IV) products.   50

Figure.3.10. DNA amplified by nested-PCR with primers U-1/MLO-7 then primers MLO-X/MLO-Y and digested with HinfI.   51

Figure.3.11. Amplification of the cytochrome oxidase (COX) sequence of DNA from dried leaves.   53

Figure.3.12. Phytoplasma in Kaui and Maui plantations sugarcane samples.   54

Figure.3.13. phytoplasma in former Hawaiian plantations sugarcane samples (2009).   56

Figure.3.14. phytoplasma in Hawaiian breeding station sugarcane samples (2010).   57

Figure.3.15. Sugarcane leaves showing symptoms of infection with yellow leaf syndrome (A) and (B), compared with an uninfected green leaf (C).   59

Figure.3.16. Phytoplasma in Hawaiian HC&Splantation of Maui island, close to Puunene (2011).   60

Figure.3.17. Phytoplasma detection in sugarcane samples from Hawaiian (Maunawili, HARC)breeding station (2011).   63

Figure.3.18. Phytoplasma identification in sugarcane samples from Hawaiian (Maunawili, HARC)breeding station (2011).   63

Figure.3.19. Phytoplasma detection in sugarcane samples from Hawaiian (Maunawili, HARC) breeding station sugarcane samples (2011).   65

Figure.3.20. Phytoplasma detection in sugarcane samples from Hawaiian (Maunawili, HARC) breeding station (2011).   65

Figure.3.21. Phytoplasma identification in sugarcane samples from Hawaiian (Maunawili, HARC)breeding station (2011).   66

Figure.3.22. Phytoplasma detection in Hawaiian sugarcane samples from different sources close to former plantations (2011).   69

Figure.3.23. Phytoplasma in Hawaiian sugarcane samples from different sources close to former plantations (2011).   69

Figure.3.24. Sugarcane leaves samples from Thailand.   71

Figure.3.25; a, b and c.Phytoplasma detection in Thiasugarcane samplesfromBang Phra and Khon Kean.   73

Figure.3.26; a, b, c and d.Phytoplasma detection in Thiasugarcane samplesfromBang Phra and Khon Kean.   74

Figure.3.27; a, b, c and d.Phytoplasma identification in Thiasugarcane samplesfromBang Phra and Khon Kean.   75

Figure.3.28;a andb.Phytoplasma in Thiasugarcane samplesfromSuphan Buri.   76

Figure.3.29. A:Standard curve.B: Log- view of standard curve chart.   78

Figure.3.30. A:Standard curve.B: Log- view of standard curve chart.   79

Figure.3.31. A:Standard curve.B: Log- view of standard curve chart.   80

Figure.3.32. A:Standard curve.B: Log- view of standard curve chart.   82

Figure.3.33. A: Thai sugarcane plant infected with sugarcane white leaf phytoplasma.   89

List of figures

Figure.3.34. Diagrammatic representation of genomic location of primers used for DNA sequencing.
90

Figure.3.35. Position of the phytoplasma strains from Hawaiian, Cuban, Egyptian and Thai sugarcane in a phylogenetic tree together with other phytoplasma isolates(Table.3.29).   95

Figure.3.36. Nested PCR results of the test plants, which received cold and hot-water treatment after 2 months post germination.   97

Figure.3.37. Nested-PCR results of test plants, which received cold and hot-water treatment after 6 months (a) and 1 year (b) post germination.
97

Figure.3.38. Nested-PCR results of test plants; which received hot water treatment at 50°C with various duration after 2 months post germination.   98

Figure.3.39. Nested-PCR results of test plants; which received hot water treatment at 50°C with various duration after 6 months post germination.   98

Figure.3.40. Phytoplasma in sugarcane source plants (containing phytoplasma), in aphids (*Melanaphis sacchari*) and in sugarcane target plants (infested by phytoplasma-infected *M. sacchari*).   100

Figure.3.41. Ultrathin section of green leaf of uninfected sugarcane (control).   102

Figure.3.42. Comparison between phloem sieve elements of white, variegated and green leaves of white leaf phytoplasma-infected sugarcane.   104

Figure.3.43. Comparison between phloem companion cells of phytoplasma infected and uninfected sugarcane.   105

Figure.3.44. Ultrathin sections of bundle sheath and mesophyll tissues of phytoplasma-infected sugarcane.
106

Figure.3.45. Transmission electron micrograph of typical membrane-bound phytoplasma bodies which present in sieve tube and contain resembling DNA in sieve tube of white leaf phytoplasma-infected sugarcane leaf.   107

Figure.3.46. Ultrathin sections of phloem tissue of phytoplasma-infected sugarcane.   107

Figure.3.47. Ultra structural comparison between paranchymatic bundle sheath cells of infected and uninfected sugarcane.   108

Figure.3.48. Comparison between chloroplasts structure of mesophyll cells in infected and uninfected sugarcane.   109

4. Discussion   110

Figure.4.1. A: Log- view of standard curve chart.   113

Figure.4.2. Binding sites of qPCR primers and probe on the nucleotide of partial 16S rRNA gene sequence of SCGS phytoplasma infects Egyptian sugarcane cultivar (G8447).   114

10. List of tables

2. Material and Methods — 17

Table.2.1. Ingredients of CTAB extraction buffer. — 18

Table.2.2. Sequences of universal primers used in the amplification of phytoplasma 16S rRNA operon. — 21

Table.2.3. Oligonucleotide primers used for nested-PCR assay I. — 27

Table.2.4. Oligonucleotide primers used for nested-PCR assay II. — 27

Table.2.5. Oligonucleotide primers used for PCR assay III. — 28

Table.2.6. Oligonucleotide primers used for PCR assay IV. — 28

Table.2.7. Sequence of primers and probes used for detection of phytoplasma and plant DNA (Christensen et al., 2004). — 37

3. Results — 39

Table.3.1. Original sources of sugarcane samples. — 41

Table.3.2. Results of nested-PCR assay (I) and identification of phytoplasmas based on RFLP analyses. — 46

Table.3.3. Phytoplasma in Hawaiian sugarcane samples. — 51

Table.3.4. Phytoplasma in Egyptian and Syrian sugarcane samples. — 52

Table.3.5. Phytoplasma in Hawaiian plantations sugarcane samples (2009). — 55

Table.3.6. Phytoplasma in sugarcane samples from former Hawaiian plantations (2009). — 57

Table.3.7. Phytoplasma in sugarcane plants from Hawaiian (Maunawili, HARC) breeding station (2010). — 58

Table.3.8. Phytoplasma in sugarcane plants from Hawaiian HC&S plantation of Maui island, close to Puunene (2011). — 60

Table.3.9. Phytoplasma in Hawaiian (Maunawili, HARC) breeding station sugarcane samples (2011). — 62

Table.3.10. Phytoplasma in sugarcane samples from Hawaiian (Maunawili, HARC) breeding station (2011). — 64

Table.3.11. Presence of SCYLV in sugarcane cultivars in the HARC breeding station in 2001 and presence of phytoplasma in these cultivars which collected in (2011). — 66

Table.3.12. Phytoplasma in Hawaiian sugarcane samples from different sources close to former plantations (2011). — 68

Table.3.13. Outline the detection of phytoplasmas; in samples were obtained from grass weeds from Sumida water cress farm in pearlidge; based on nested-PCR assays. — 70

Table.3.14. Phytoplasmas in sugarcane samples from provinces of Bang Phra and Khon Kean in Thailand in 2010; based on nested-PCR assays. — 72

Table.3.15. Outlines of phytoplasmas in Thai sugarcane samples from Suphan Buri. — 76

List of tables

Table.3.16. Performance characteristics of qPCR assay. 83

Table.3.17. Q-PCR results from 10-fold dilution series of the reference samples which diluted in water or in healthy host plant DNA. 83

Table.3.18.Q-PCR results of the Hawaiian sugarcane samples grown in greenhouse. 84

Table.3.19.Q-PCR results of the Egyptian and Syrian sugarcane samples. 84

Table.3.20.Q-PCR results of the Hawaiian plantations sugarcane samples (2009). 84

Table.3.21.Q-PCR results of the Hawaiian breeding station sugarcane samples (2010). 85

Table.3.22.Q-PCR results of the Hawaiian grass weeds from Sumida watercress farm (2011). 85

Table.3.23.Q-PCR results of the Hawaiian sugarcane samples from HC&S, Maui, close to Puunene (2011). 85

Table.3.24.Q-PCR results of the Hawaiian sugarcane samples from HARC in Maunawili (2011). 86

Table.3.25.Q-PCR results of the Hawaiian plantations sugarcane samples from different sources, (2011). 86

Table.3.26.Q-PCR results of the Thai sugarcane samples from Bang Phra and Khon Kean provinces. 87

Table.3.27.Q-PCR results of the Thai sugarcane samples from Suphan Buri province. 87

Table.3.28.Q-PCR results of the distribution of the phytoplasma in sugarcane plant. 88

Table.3.29.Phytoplasma strains and their GenBank accession numbers used in this study for the phylogenetic trees (Figure.3.35). 92

Table.3.30. Effect of hot water treatment on phytoplasma elimination of cultivars H65-7052 and H78-77 50 according to the indicated temperature regime. 99

Table.3.31. Q-PCR results of the insecticide treatment of phytoplasma-infected sugarcane plants. 99

i want morebooks!

Buy your books fast and straightforward online - at one of world's fastest growing online book stores! Environmentally sound due to Print-on-Demand technologies.

Buy your books online at
www.get-morebooks.com

Kaufen Sie Ihre Bücher schnell und unkompliziert online – auf einer der am schnellsten wachsenden Buchhandelsplattformen weltweit! Dank Print-On-Demand umwelt- und ressourcenschonend produziert.

Bücher schneller online kaufen
www.morebooks.de

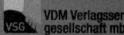 VDM Verlagsservicegesellschaft mbH
Heinrich-Böcking-Str. 6-8   Telefon: +49 681 3720 174   info@vdm-vsg.de
D - 66121 Saarbrücken   Telefax: +49 681 3720 1749   www.vdm-vsg.de

Printed by Books on Demand GmbH, Norderstedt / Germany